HANS KREBS: REMINISCENCES AND REFLECTIONS

HANS KREBS

REMINISCENCES AND REFLECTIONS

In collaboration with
Anne Martin

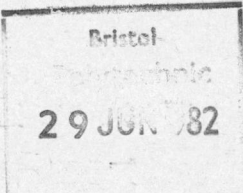

CLARENDON PRESS OXFORD 1981

Oxford University Press, Walton Street, Oxford OX2 6DP
OXFORD LONDON GLASGOW
NEW YORK TORONTO MELBOURNE WELLINGTON
KUALA LUMPUR SINGAPORE JAKARTA HONG KONG TOKYO
DELHI BOMBAY CALCUTTA MADRAS KARACHI
NAIROBI DAR ES SALAAM CAPE TOWN

© Hans Krebs 1981

All rights reserved. No part of this publication may be reproduced, stored in a retrieval system, or transmitted, in any form or by any means, electronic, mechanical, photocopying, recording, or otherwise, without the prior permission of Oxford University Press

British Library Cataloguing in Publication Data
Krebs, Hans
  Reminiscences and reflections
  1. Krebs, Hans
  2. Scientists – England – Biography
  I. Title   II. Martin, Anne
  509'.2'4   Q143.K/
  ISBN 0-19-854702-1

Reproduced from copy supplied
printed and bound in Great Britain
by Billing and Sons Limited
Guildford, London, Oxford, Worcester

PREFACE

In 1960 I was asked by the Committee of the British Biochemical Society to deliver the Annual Lecture in memory of Sir Frederick Gowland Hopkins, the foremost figure in British biochemistry during the first three decades of this century. Most of this lecture was concerned with biochemical matters, but in the final section I spoke of personal things—of the impact upon myself of Hopkins and his laboratory, where I found refuge in 1933 after Hitler's Germany had barred my career in the country of my birth. This lecture was published in the *Biochemical Journal* [182] and many who read it expressed keen interest in the story I told (quoted on pp. 92–3 of this book). Indeed it seemed to arouse rather wider interest than the scientific part and readers encouraged me to write more about my personal experiences. That was the starting point of this book.

At that time I was very busy with my day-to-day work of teaching, research, and administration, but the idea took shape that I might put together for publication some documents from my files—those that had been important in influencing the course of my professional life and those that reflected more personal aspects, such as letters, announcements, and published articles. In 1969 I showed some of this material to the Oxford University Press and asked whether publication seemed worth while. The answer was a request for a considerable expansion to include a fuller account of my personal life.

To begin with I had reservations. I thought that the documents would speak for themselves, illustrating how the social and political history of the times was reflected in the career of an individual scientist. Among other things they showed how the recognition of research expresses itself, how the rise of Nazism temporarily upset my career, and how British scientists and the Rockefeller Foundation came to my rescue; and how, later on, I became involved in wider issues. The request from the Oxford University Press and the encouragement of friends and colleagues overcame my inhibitions and so I have extended the text by recounting and commenting on some of my experiences. I have, however, limited myself essentially to those experiences which directly concerned my life as a scientist. I make no more than passing reference to my family life and personal friendships. Both have been important to me but, since they had no direct bearing on my scientific career, I felt that they would be of little interest to readers.

I have never kept a diary, except for a few periods, such as during the war years and on some of my travels. But throughout my adult life I have retained documents and correspondence which I thought I might like to read again or which might be of interest to my family. These documents have provided the backbone of this book.

Much has been said and written about the pitfalls of autobiographical writing (1–7). George Bernard Shaw (1) wrote, in characteristically provocative manner:

All autobiographies are lies. I do not mean unconscious, unintentional lies: I mean deliberate lies. No man is thought to be good enough to tell the truth about himself during his lifetime, involving, as it means, the truth about his family, his friends and his colleagues. And no man is good enough to leave the truth to posterity in an autobiography which he suppresses until there is nobody left alive to contradict him.

In the same book, tongue in cheek he commented that 99.5 per cent of his life was like that of the rest of us and the remaining 0.5 per cent that was different could be seen on the stage and read in his books.

I could say something similar myself.

Goethe (2) said,

The question of whether or not one should write one's own biography is inappropriate. Those who do it I consider the most polite of all people. If a person wishes to write about himself his motives are irrelevant. It is not at all necessary that he is a virtuous person and that his deeds are distinguished and virtuous. All that matters is that something is done which may be of use and give pleasure to others.

Peter Sloterdijk (8) has recently discussed to what extent autobiographies are literature, i.e. pieces of art, and to what extent merely an account of experiences. He comes to the conclusion that autobiographies 'burst open at the frame of literature as art'. I cannot make a claim that my writing is a

PREFACE

piece of art; it is more a piece of science and its style therefore aims at clarity and simplicity. Thus if it has a value it is not as a piece of literature but as a record of the experiences of one man's life.

I have addressed myself to a general readership, but because some of the recollections and reflections recorded here refer to my life as a scientist, the technicalities of my science cannot be entirely omitted. However, readers may bypass these passages without losing the continuity of the story.

It has been my intention—following my habit as a scientist—to make the book a scholarly one by giving sources of my information, and some additional notes of interest. These are given as appendices so that they do not interrupt the flow of the text.

The editorial staff at the Oxford University Press were most helpful in offering criticisms and advice. I also wish to record my thanks for the help of Helmut Sies (Düsseldorf), Frederic L. Holmes (Yale University), Stephen Kearsey (Cambridge University), and Margaret, my wife.

My very special thanks are due to Anne Martin, who has been my close collaborator from 1978 to 1980. By asking an endless number of challenging questions—What did I do? What did I feel? What did I expect? What did I believe? What were my reactions?—she made me think about aspects which otherwise may not have crossed my mind. By criticizing what I had written—style and subject matter, and by revising the text, she has introduced many improvements. By her hard, reliable, careful work, her enthusiasm and willingness to help, her always cheerful mood, and her sustained efforts, she has been an ever-refreshing stimulus.

Oxford
December 1980
HAK

CONTENTS

	List of plates	x
1	Beginnings (1900–1918)	1
2	University (1919–1925)	16
3	In the laboratory of Otto Warburg (1926–1930)	27
4	Altona and Freiburg (1930–1933)	40
5	The discovery of the ornithine cycle of urea synthesis (1932)	51
6	Hitler's seizure of power (1933)	61
7	Cambridge (1933–1935)	83
8	Settling at Sheffield (1935)	95
9	The history of the discovery of the citric acid cycle (1936–1937)	105
10	The war years (1939–1945)	119
11	The history of the discovery of carbon dioxide fixation in animal tissues	126
12	The Medical Research Council Unit (1944–1967)	132
13	Re-establishing contacts with Germany (1945–1949)	147
14	Final formulation of the citric acid cycle	158
15	The Nobel Prize (1953)	165
16	Academic invitations (1946–1954)	180
17	Oxford (1954–1967)	193
18	'Retirement'	222
19	Envoi	230
	Notes	233
	People in the book	247
	References	263
	List of publications	269
	Index	291

LIST OF PLATES

(Plates fall between pp. 146 and 147)

1. The author's parents.
2. Hildesheim.
3. (a) The author as a student about 1922; (b) at the Congress of Internal Medicine 1932.
4. F. G. Hopkins and T. Thunberg.
5. Nobel Prize-giving ceremony.
6. Lecturing at Lindau, 1966.
7. The author with his team, 1974.
8. The author with Estabrook, Ochoa, Bloch, and Cori, 1980.

NOTES

Numbers in parentheses refer to the list of references, p. 263; numbers in square brackets refer to the author's list of publications, p. 269; and superior numbers refer to the notes, p. 233.

I
BEGINNINGS
(1900–1918)

On 20 June 1933 I found myself on board the Channel Ferry, bound for England. I did not think my enforced emigration from Nazi Germany would be a long one: I did not expect the Third Reich to last very long, because I believed that the victors of the First World War would not wish to abandon the fruits of victory—peace—and that they would be wise enough to take appropriate action. At the beginning of the Nazi regime Germany was virtually unarmed and the ousting of Hitler and his warmongers would have been very easy for the Western powers. Jokingly we said, 'We will patiently await the end of the Thousand Years of the Third Reich', which Goebbels, Minister of Propaganda, had proclaimed.

Alas, this turned out to be wishful thinking, but at the time it alleviated the pain of leaving my homeland. Not foreseeing the course of future events I was not unduly upset that I was leaving many close relations and friends behind me. There did not appear to be an immediate threat to their lives or livelihoods since none had been politically active.

I crossed the German frontier at Strasbourg and felt enormously relieved when I passed customs and passport control and reached French soil. Sixteen eventful years were to pass before I once more set foot in Germany, but I was never to see again the town I knew and loved as a child for it was to be completely destroyed by bombs on 22 March 1945, two days before the Allied armies were to force the crossing of the Rhine.

Hildesheim, where I was born on 25 August 1900, the son of George and Alma Krebs, was then a quiet town of some 50 000 inhabitants. It lay in North Germany, near Hanover. It was a uniquely beautiful town and had a long and distinguished history. From around the tenth century until the Thirty Years' War in the seventeenth century, Hildesheim had been one of Europe's major cultural centres. In the old quarter of the town, the greater part of which was surrounded by earthworks and moats (in places as wide as the Thames at Windsor) there were innumerable buildings to bear witness to Hildesheim's great past.

While still a boy I was impressed by the beauty of the medieval town, essentially unspoiled by architecture of later centuries. Magnificent churches contained art treasures of international fame; the great public buildings and the luxurious patrician houses of wealthy citizens were richly decorated with paintings and carvings. Many bore inscriptions (1) expressing a variety of sentiments—piety, wit, humour, the appreciation of beauty and pride. '*Spero invidiam*', said one, 'I hope to be envied'; and another, '*Unsere Vorfahren waren auch keine Narren!*' 'Nor were our forebears fools'.

Clustered around these were hundreds upon hundreds of more humble, half-timbered dwellings with warm red roofs. Though all were built in the same style, no two were exactly alike. Their individuality, and the charm of the narrow, winding cobbled streets, conveyed a sense of character and vitality. The shapes of the houses and the texture and pattern of their surfaces were of a tasteful simplicity, an ageless character, a kind of beauty to which the half-timbered style lent itself readily. Intermingled with the buildings was a good deal of greenery. Despite the individuality of the buildings, the general impression of the old town was that of an harmonious whole. There were no demonstrative expressions of extravagant, ephemeral tastes, and of course no disfiguring advertising. Tradesmen and craftsmen advertised their goods and skills discreetly, in a dignified manner. It was evident that the planning of buildings and streets was not dominated by purely utilitarian considerations, but was the result of a love of beauty as well as of personal and civic pride.

The long tradition which Hildesheim reflected evoked in me a sense of enrichment, a feeling that I possessed something that did not exist elsewhere. It also evoked a sympathy with, and respect for, the town's earlier generations. To be the youngest link in the chain of its traditions was a source of pride and inspiration.

The old town and its charming background of gently rolling wooded hills were a constant source of delight and attraction to me. I would roam around, on my bicycle or on foot, mostly alone, exploring back streets and hidden corners, every hill and wood, savouring familiar prospects and seeking out new ones.

BEGINNINGS (1900–1918)

During my travels later on in life I have seen many beautiful towns, but none has matched Hildesheim's distinctive charm. I have never lost my love for it although my enforced emigration greatly strained my emotional attachment and disrupted my physical connection with it for sixteen years. Lest it be thought that my description of Hildesheim is coloured by parochial affection for the place in which I spent a happy youth, I would like to quote a few impressions which a complete stranger, E. del Monte, a visitor from France, recorded in 1883 (2).

It is like a dream, a vision, to walk through the maze of streets, each more remarkable than its fellows, yet none damaging the harmony of the whole. All around are tableaux such as are not to be seen except in Venice or Nuremberg. As with the whole, each tiny detail, too, is of unusual significance.

The rise and fall of the ground and the twisting of the streets unfold a series of picturesque surprises. The irregularities of the houses and the brick walls of the gardens retain the town's aspect of bygone days, despite the pavements and gas lamps. Here and there one finds places planted with limes or acacias, whose foliage blends most marvellously with the roofs of lively red—so steep that many are taller than the house walls.

It is to this wondrous *mélange* that the town owes its atmosphere of gaiety—even in the rain. It is a triumph of colour, a triumph lacking in Paris where the sandstone—the horrible yellowy-grey muddy sandstone—cries out for something to enliven it.

Most of the houses are skewed with age: where the ground has sunk the projecting upper storeys have tilted forward—in some streets they seem to be trying to bend closer to salute each other.

The townspeople are full of tender care for these silent witnesses of the past, and full of concern that they shall be preserved for as long as possible.

People in Hildesheim and other places that were destroyed by bombing so near to the end of the war, often refer to 'those senseless acts'. Do they not appreciate that as late as 19 March 1945 (three days before the final heavy raid on Hildesheim) Hitler issued an order that, 'All military, transportation, communications, industrial and food supply facilities are to be destroyed, along with anything else within the Reich that might in any way, now or in the future, be of some use to the enemy in the continuation of the struggle'? Those who did not comply were to be shot (3).

When Albert Speer, then responsible for the whole German war economy, tried to remonstrate, Hitler replied icily, 'If the war is lost, the nation will perish. There is no need to consider the resources people would need, even for their most primitive survival. On the contrary, it is better that we destroy all these things ourselves. For the nation will then have shown itself to be weak and the future to belong totally to the stronger eastern peoples. In any case, those who are left at the end of the struggle are of little worth, for the good have fallen'.[1] As Sebastian Haffner put it, 'Die Vernichtung Deutschlands war das letzte Ziel, das Hitler sich setzte' (The annihilation of Germany was the last goal Hitler set himself).

An account (4) of the Royal Air Force bombing policy states: 'Naturally the armies expected the opposition to be very heavy, for would not the Wehrmacht be defending the sacred soil of the Fatherland? They therefore urged strongly that the bombing programme should begin several days before the crossing [of the Rhine], which had been fixed for 24 March.' The British commanders were aware of the possible destruction of great art treasures, but acted on the principle that saving such treasures could not be at the expense of the lives of Allied soldiers. Any blame would rest with the German commanders who, obeying Hitler's decree, were prepared to fight to the last. The official British comment on the raids on Dresden on 15 February 1945 was: 'Dresden, the German Florence and the loveliest rococo city in Europe, has ceased to exist.'

Today, Hildesheim is a modern, busy town which has grown to over 100 000 inhabitants. The few great churches and other buildings which survived the air raids or were later restored still bear testimony to the town's history, but they are mere remnants of its former magnificence and splendour. Lacking the frame of their original medieval setting, flanked instead by modern buildings in widened streets, they now seem rather like museum pieces and no longer diffuse the historical traditions of the town in the lively way they did.

Our family home was close to the old town, but relatively modern, having been built in 1870. My parents moved there with my five-year-old sister Elisabeth (called Lise or Lisa) when I was six weeks old. My brother Wolfgang (Wolf) was born there two years later.

The house was quite large, surrounded by a good-sized garden. As well as being our home it also contained my father's consulting rooms. He was a successful and much respected surgeon who specialized in diseases of the ear, nose, and throat. In his student days he had hoped to be able to follow a university career, but this would have meant prolonged financial dependence on his father. So, at the age of twenty-five, when he had completed his specialist training, he decided to set up in practice. The development of otolaryngology remained a lifelong interest and between the years 1895 and 1914 he contributed articles based on his clinical experiences to the medical literature at the rate of one or two a year.

My father had been born and brought up in Silesia, but chose to settle in Hildesheim. He was attracted by the beauty of the town and its surroundings and also by the fact that his speciality, which was only just emerging, was not yet represented there.

He was a man of wide and profound intellectual interests. He read a great deal and was an enthusiastic theatre- and concert-goer. He was an excellent, if not very frequent, letter-writer, and when inspired by emotion could express his feelings in elegant poetry—for example, in the form of odes to his wife. At family celebrations and other social events he would contribute to the entertainment with amusing and polished light verse—a custom which had been common in Germany since the eighteenth century. Altogether he created in the home a stimulating, intellectual atmosphere which we three children accepted as normal.

He took a keen interest in politics; he was a convinced democrat and German patriot. He came to believe that the only answer to the problem of anti-Semitism was the complete assimilation of all Jews into the community in which they lived, which included abandoning their Jewish faith. This was also the policy encouraged by the government of Imperial Germany.

In my childhood I never encountered anti-Semitism personally though I knew very well that it existed. In Germany before 1919 it had not the viciousness which was to appear after Hitler founded his National Socialist party in 1920; it showed itself by the exclusion of professing Jews from senior posts in the Civil Service and from the Officer class. A few hotels and holiday resorts made it unofficially known that Jews were not 'welcome'. The promotion of Jews to senior posts was rare and much pressure was put on potential candidates to be baptized. This often removed the bar.[2]

There was an anti-Semitic gutter press and a number of widely-read books (5). One of these was by Adolph Stoecker, chaplain to the Kaiser and a politician who founded a specifically anti-Semitic political party. His views were based on religious and nationalistic beliefs and his activities were directed especially against the liberal policies of the Social Democrats. Another book which was very popular in Germany was by Houston Stewart Chamberlain, a British writer who styled himself 'cultural philosopher' and who believed in the superiority of the 'Nordic' races. A more academic book which I studied in my teens was by Eugen Dühring.

The reactions of German Jews varied. The Zionists' answer was a return to the Promised Land; others preferred to hold to tradition and to demonstrate this openly; a third group, to which my father belonged, believed that the only real answer lay in assimilation.

Believing as he did, he went so far as to deny his sons any Jewish religious teaching. But since some kind of religious instruction was then compulsory in Germany, Wolf and I attended scripture lessons in school. These, with their Protestant accent on the Bible, gave me a thorough introduction to the scriptures, an experience which I have always treasured. In the same spirit of assimilation, my father also chose for his children what he believed to be truly Germanic names, Elisabeth and Hans. Years later he was to be mildly amused when a friend told him that both names were of biblical, and hence Hebrew, origin. He made a happier choice, from his point of view, in calling my younger brother 'Wolfgang', after his favourite poet, Goethe, and favourite composer, Mozart.

My mother was a quiet person. She was devoted to her husband and children and happy to centre her whole existence upon them; she cared for us unfailingly. When not occupied with household duties, she spent her time reading, sewing, or visiting her many relatives. She shared many of my father's interests but was a passive companion rather than an active contributor. On my own intellectual development she had no major influence. Born Alma Davidson, she was a native of Hildesheim. Her ancestors had lived in the district for centuries

BEGINNINGS (1900–1918)

and we had more than a hundred relatives living near by. Today only one is left in Hildesheim. In the Nazi holocaust the family became scattered all over the world; not a few perished in concentration camps.

Most of my mother's relatives were in business except for one, her cousin Isidor Traube, who was a university professor. He was a physical chemist at the Technical University of Berlin-Charlottenburg. In 1934, at the age of seventy-four, he came to Britain as a refugee. He was welcomed as a guest by the University of Edinburgh and continued his research there for some years. He is still mentioned in textbooks as the discoverer of 'Traube's surface tension rule'.

Our home life was simple, indeed quite Spartan if compared to the comfort that the average family enjoys today. Our everyday diet was only marginally adequate and we had to eat everything that was served, whether we liked it or not. The food was plain: bread, vegetables, and milk were the staple items; with cheap margarine rather than butter; and never jam and margarine on our bread, only one or the other. Cakes and puddings appeared only on Sundays and birthdays, while chocolates and sweets were rare treats. Toys were very simple and Christmas and birthday presents little more than necessities such as new clothes. Visits to a cousin who had a collection of Meccano and an elaborate model railway were very popular with Wolf and myself.

This austerity was the accepted way of life for many middle-class German families in the early part of the century. It was not enforced by financial stringency; the frugality was a matter of thrift, of saving for what were regarded as essentials as well as for lean days and old age. Essentials, to my parents, included a good education for their children, leading perhaps to university and the training for a profession. This depended, for most young people, entirely on financial support from the family.

Both our parents tended to be stern with us and to expect high standards in every sphere of conduct, behaviour, school work, and personal discipline. Sometimes I chafed when I thought my mother was 'nagging' me, but perhaps she had found this the only way she could make me behave as she felt I should. Any demonstration of emotion was frowned upon; there were no spontaneous hugs or goodnight kisses. The 'stiff upper lip' characteristic of the ruling classes (the officer class in particular) set the tone throughout most of northern Germany in those days.

It was my father who first aroused my interest in living things. He had a love for the countryside, especially its plants and birds, and he took us for walks in the hills. The beauty of the country and its wild life made a deep and lasting impression on me. To this day I can relive the experience, when on one of our family outings, I found a rare orchid—*Cypripedium calceolus* (the lady slipper)—and even now could point to the spot where it was growing. I remember where I saw carpets of *Corydalis cava* both white and purple (not found wild in Britain; it is a relative of the yellow fumatory) interspersed with *Anemone nemorosa* (wood anemone) and *Primula vulgaris* (cowslip). Elsewhere, in their season I recall profusions of *Vicia silvatica* (wood vetch), *Helleborus viridis* (green hellebore), and *Convallaria majalis* (lily-of-the-valley). A few of the wild flowers abounding around Hildesheim now grow in my garden at Oxford, from seeds which I collected some years ago.

Since my parents were eager readers and subscribed to a book circle, there were always plenty of books around the house. I liked browsing through encyclopaedias and other reference books. Indeed I had a great thirst for knowledge of all kinds. My father sometimes chided me for indiscriminate reading, for absorbing facts without digesting them. When I was nine or ten I wanted to find out what was going on in the world and began to read newspapers. Not all my reading was erudite: comic and nonsense verse, such as that by Christian Morgenstern, could keep me chuckling for hours. Not only did I enjoy reading, from quite an early age I had a desire to own books, to see them around me, to be reminded of them and to be able to re-read them. Alas, not until I was much older, in my thirties, could I afford to buy more than the occasional one.

Reading and going to theatres and concerts were the highlights of my youth. Hildesheim had its own theatre even though it was a relatively small town. The theatre was heavily subsidized by the City Fathers as was, and still is today, common practice in Germany. The theatre is looked upon as an important civilizing influence, worthy of substantial municipal support, because of the conviction that, while providing good entertainment, the theatre opens the way to the soul and genius of mankind throughout the ages and is a powerful supplement to literature and the arts.

Walking in the wooded hills around Hildesheim and cycling were our main outdoor activities. Cycling had to stop in 1916 or 1917 because of the shortage of rubber in beleaguered Germany, so I took to canoeing on the near-by river. Sometimes we played tennis on a court which belonged to a cousin and therefore cost us nothing.

Holidays away from home were rare. Only twice during my childhood did we take a family holiday further afield—to the Dutch seaside at Scheveningen in 1909 and to Oberstdorf in the Bavarian Alps in 1911. Both of these holidays left a deep impression on me; the beauty of the Alps has remained a life-long attraction. I also remember vividly the visits to museums in Amsterdam, in The Hague, and in Munich *en route* to our holiday destinations.

Lise, Wolf, and myself were, on the whole, contented children. Lise, five years my senior, was a gentle girl and liked to 'mother' her two younger brothers. Wolf was the closest childhood playmate I had. All three of us suffered from being shy, especially when in strange company. I can remember my father spanking me on one occasion before I could be persuaded to enter a strange house to deliver a letter. My sister tells me that even now, in her eighties, she still has not overcome her basic shyness. My own disappeared only very gradually as confidence grew from being accepted by people whom I respected.

This account of my childhood may give the impression that I was a model boy. Neither I nor my family thought so. True, I was diligent and conscientious in my schoolwork and I tended to enjoy serious pastimes. I was self-conscious, timid, and solitary and therefore never aggressive nor rebellious—on the contrary I was anxious to conform. Both family and school criticized me for being untidy and my parents blamed me for being domineering towards my brother. So I was made aware of my shortcomings. Altogether I formed the impression that I was a rather unattractive and unpopular boy.

The local grammar school, the 'Andreanum', which I attended from 1910 until 1918, dates back to the beginning of the thirteenth century. The earliest documentary evidence is dated November 1225, when Archbishop Siegfried II of Mainz removed the restrictions which had limited to forty the intake of pupils at the school attached to St Andrew's Church. Teaching had in fact begun there in 1203 but for the purposes of official celebration the year 1225 was adopted as the founding date. When the school commemorated its 750th anniversary on St Andrew's Day in 1975, I was asked to take part in the celebrations and to give a speech. The school orchestra also contributed to the occasion and played the music of another ex-pupil, Georg Philip Telemann. Some of the pieces selected were taken from his 'Musical Geography' Suite, composed whilst he was a pupil at the school from 1697 to 1701.

At times the school had achieved great distinction, but a long history is no guarantee of continuing high standards. While I was there the Andreanum was of rather average quality, particularly after 1914 when the war called away almost all the younger masters. It was a *Humanistisches Gymnasium*, concentrating primarily upon Latin, Greek, History, and German. Mathematics were well taught but the other sciences were neglected. Instruction in biology, physics, and chemistry was very sketchy, so that when I went to university to read medicine, I had to start these subjects from scratch. But I have never regretted my classical education. It revealed to me the depth and quality of the Greek and Roman civilizations, and could not have been replaced later in life. It helped me to see the modern world in perspective and gave me a deeper understanding and appreciation of the languages we speak today. Moreover, it drove home the fact that in some areas of human endeavour, the achievements of the Greek and Roman civilizations have never been surpassed.

All the subjects interested me, although the depth of my interest varied with the quality of the teacher. I was a good pupil but not outstanding in any subject; it never occurred to me, or to anybody else, that I had any special potential. One of my favourite subjects was history. Once, during the German course, we were asked to write an essay on a subject of our own choice. The subject I chose (after discussing it with my father) was 'On whom does history confer the title, "The Great"?' I had in mind in the first instance Alexander, Charlemagne, and Frederick. On searching through history books I found only a few more, among them Alfred. Although the essay earned me good marks it was, of course, rather superficial, as I realized at the time. But it was a beginning and helped to prepare me for a few historical writings [274, 276, 278, 288, 294, 296, 301, 311] which I undertook later in my life.

Latin I found enthralling after a superb teacher introduced us to Horace; I committed quite a number of the poems to memory. Another experience which has never faded from my mind was reading the defence of Socrates, in Plato's *Apology* (6). I was struck by Socrates' awareness of his own lack of knowledge, as evinced in his statement, 'Is not the most reprehensible form of ignorance that of thinking one knows what one does not know? I differ from other men in this way, and if I were to say that I am wiser in anything, it would be in this: that not knowing, I do not think I know.'

On music I spent more time (in hours of practice) than I did on any one subject. I never managed to become an accomplished player, but throughout my life music has given me deep pleasure and relaxation through both listening and playing, however imperfectly.

My piano teacher, Richard Gerlt, had a strong educational influence upon me, especially upon my standards of workmanship. He was a very able teacher and concert pianist. He patiently taught me to play precisely according to the composer's instructions. He would not tolerate slipshod playing and so I had to accept a great deal of criticism, but I never resented it because he was right and because he was friendly, warm-hearted, and kind.

He did not talk to me only of music and musicians. Sometimes he would voice his thoughts on the everyday affairs of life. I recall, for example, a comment he made about concern with one's outward appearance. Was it, he asked, all a matter of vanity? He thought not; he saw it also as a matter of aesthetics. Neglecting one's appearance might hurt aesthetic sensibilities, while paying attention to it might give pleasure. Since I have never forgotten this observation of his, it must have impressed me very much. By touching frequently, though briefly, on matters of this kind, he contributed to my general education. I was too shy to take much part in our conversations, but listened with awe to all that he had to say.

Playing the piano (or rather, trying to play) taught me,

amongst other things, that no matter how hard one might try, the standards that can be achieved in technique and in interpretation vary from person to person. This means that innate talent, i.e. genetic factors and individual differences in these genetic factors, sets a limit on the proficiency which can be attained in any field. It has also taught me that even those who are gifted can reach and maintain their peak of performance only by very hard work.

But hard work and ambition are not enough. This was a lesson for life. Much later, when students discussed with me their future careers, it was very much in my mind that success and happiness depend on choosing a career most suited to the given talents. Some bright students at Oxford express the view that almost anybody with a modicum of intelligence can be successful if he works hard. The genius, they say, is successful without working hard; they call this 'effortless superiority' (7).[3] I have never come across this 'effortless superiority', either in science or anywhere else.

I had little difficulty in passing school examinations, usually coming sixth or seventh in a class of about twenty-five. Nevertheless, my father often gave me to understand that he was sceptical about my intellectual potential. He allowed that I managed to absorb facts readily on most subjects and that I could cope with mathematics, but he criticized me for not having developed a philosophy of life (*Weltanschauung*). Other boys of my age, he told me, were often much more mature and had begun to form their own opinions on various aspects of life. He often quoted to his children, with an air of resignation, what he thought to be a Latin proverb,[4] 'You cannot make a golden Mercury out of wood' which corresponds to the English saying, 'You cannot make a silk purse out of a sow's ear'.

My father's critical comments helped me to be modest in assessing my abilities and potential, but they did not discourage me unduly. My wish to follow in his footsteps matured very early, at the age of fifteen. I was impressed by his work, and could see from the frequent expressions of gratitude he received from his patients that a doctor's life could be a gratifying one. That the life would be strenuous, with not infrequent night and Sunday calls or even an interrupted holiday, was something I took for granted and was in no way a deterrent.

BEGINNINGS (1900–1918)

My early inclinations to follow in my father's footsteps as a surgeon were much encouraged when reading the *Iliad* in Greek at school I came across the words (8): ἰατρὸς γὰς ἀνὴς πολλῶν ἀντάξιος ἄλλων (a doctor is not one life, but the lives of many). This must have impressed me because it is one of the few passages from Homer which imprinted itself indelibly into my memory although it was not one we were required to commit to memory. It occurs in the following sentence: 'When Machaon the field surgeon was wounded, the warriors called upon Nestor to take Machaon quickly to safety, for a doctor is not one life, but the lives of many.'

My father often reminded Wolf and myself that earning a living would be difficult because of the severe competition for customers—whether one became a doctor or a cobbler. He warned us that we would have to be very competent and reliable in our chosen field.

The preparation for a medical career meant that I would not be financially independent for many years. In the event, I was twenty-five years of age before I obtained my first paid position, as assistant to Otto Warburg. I was very proud that at last I could support myself; so pleased, that in a letter to my father telling him of my appointment, I added that should he ever be in difficult financial straits, I would be glad to help him! I knew, of course, that he was comfortably off, and did not expect to be taken too seriously. However, when currency restrictions would have prevented his visiting me in England in 1935 and 1937, and after the Nazi regime deprived him of the right to earn a living in 1938, these sad and unforeseen circumstances gave me the opportunity to help him, in a modest way.

Shortly before my fourteenth birthday, the First World War began. Neither I nor those around me had any inkling of the disastrous consequences it was to bring. According to H. A. L. Fisher (9), this failure to visualize the future was world-wide. We had witnessed the Italian–Turkish war of 1911 which ended with Italy's conquest of Libya. We had witnessed the two Balkan wars of 1912 and 1913. All of these had been short, and in 1914 we expected another speedy conclusion and were confident of victory. Only one person I knew, an elderly uncle in the cloth trade, foresaw the tragedy, but his clearly expressed pessimism did not impress us. No-one in public life expressed, or appeared to have, any doubts—except some deeply committed pacifists, and they were vilified as traitors by the authorities and the press. At least one responsible politician in Britain had some forebodings, though he did not express them in public. This was the Foreign Secretary, Sir Edward Grey, who, on the day of Britain's entry into the war, said to a friend, 'The lamps are going out all over Europe; we shall not see them lit again in our lifetime' (10).

During the first year of the war, there was little change in the everyday life of Hildesheim, though men of military age, up to 40, disappeared and the news of the appalling numbers of casualties was perturbing and depressing. As the war progressed, goods became scarce and food rationing was introduced. I started to grow some vegetables and raise some poultry in our garden, and so found another life-long hobby.

My school career was ended six months prematurely by my conscription into the army in September 1918. Those who were called up were allowed to sit an 'emergency' higher school-leaving examination; my marks were so good that I thought the examiners had been unduly lenient and sympathetic—as they tended to be with those who were going to war.

My service in the German army, with a signals regiment in Hanover, was of short duration. It came to a sudden end two months later, with the military and political collapse of Germany. This collapse came to me and my fellow soldiers as something of a shock. The strict censorship of the press had kept us in the dark about the realities of the situation, and we seldom had serious discussions. At the time I joined up I did not know, though I read the papers regularly, that Germany's allies, Turkey and Bulgaria, had already collapsed, or that Austria had put out peace feelers to the Allies. Nor did I know that the German High Command, after the major victories of the Allied offensive on 8 August, had fully recognized the hopelessness of the military situation and had instructed the German Government to sue for an armistice.

In retrospect, all this seems amazing. Today I am still puzzled at the fatalistic way in which my generation, as young men of seventeen or eighteen, accepted our conscription at that stage of the war. It should have been obvious to all clear-thinking people, in spite of the censorship, that our prospects of avoiding defeat, let alone of winning the war, were minimal and that to continue to fight was useless. Perhaps, caught up as we were in the efficient, disciplined war machine of the German Army, we realized that there was nothing we could do but accept our fate. Not only was the futility of our call-up not discussed amongst my fellow soldiers, it had not been discussed by my friends and teachers, nor even within the family. The only person who spoke to me about it in clear terms was Richard Gerlt. When I went for my final lesson before my call-up, I listened quietly to his guarded comments on the uselessness of continuing to fight, but I was still too diffident to do more than listen.

Extraordinarily, morale (in the sense of willingness to carry out orders) was high in the regiment right up to the end, thanks to the discipline of army life. Brief though my service was, some aspects of it made a lasting impression: the clear definition of individual responsibilities within a team; the importance of the utmost personal reliability and of self- and corporate discipline. I began to realize how such qualities are inculcated: partly by example, partly by training, and partly by clearly defined rules of action.

On the morning of 7 or 8 November we were surprised to be told by a non-commissioned officer that a revolution had broken out and that we would be given ammunition to fight it. We waited in readiness for an hour or two, until it emerged that all our officers had fled, in civilian clothes. Some of the leaders of the revolution—sailors from the naval base at Kiel who had mutinied in order to frustrate attempts to take the fleet to sea for a last, suicidal battle—arrived at the barracks. They told us to put away our rifles and join them.

There followed several days of chaos, during which most of us decided to return home. Gradually some order returned, thanks to the collaboration agreed on 10 November between the army and the new political leaders, the Social Democrats led by Friedrich Ebert and Philipp Scheidemann. Members of the armed forces were asked to apply for their formal release, which was approved as soon as one could indicate that one had a job to go to. Later the same month I returned to the barracks to claim my discharge on the grounds that I would become a university student, and this was immediately granted.

2
UNIVERSITY
(1919–1925)

After a few days at home to get ready for the life of a student, I set off on 4 December 1918 for Göttingen University. Term, of course, had already begun but those returning from the services were allowed to start at any time. I was accompanied by a friend in similar circumstances, Erich Stern; he is now practising as a physician in Decatur, Illinois.

Life in Germany remained primitive and difficult for some time. The country was exhausted. Its resources of food, transport, fuel, and many other necessities were depleted; food rations were far below the accepted nutritional standards, public transport was erratic, gas and electricity were in such short supply that people were not allowed to light their houses after ten o'clock at night. But despite these privations, the men returning from the war were intensely keen to study and they plunged into the world of learning with tremendous enthusiasm. Many had felt frustrated and intellectually starved during their period of war service and they looked forward with great expectation not only to preparing themselves for their chosen career, but also to the opportunity of experiencing university life. Those who had been students before us—our teachers and fathers—had told us how greatly they had appreciated the new freedoms they had found there, and that they looked upon their university years as highlights of their lives. So we expected a lot. My expectations were not disappointed.

What my generation at the end of the First World War, like generations before, thought of universities still held true thirty years later when it was well expressed by John Masefield (1), then Poet Laureate. On 25 June 1946 he received an honorary degree at Sheffield University and, in his reply to the toast of the Honorary Graduands, we heard him say:

There are few earthly honours more to be prized than this which you are now giving us.

UNIVERSITY (1919–1925)

There are few earthly things more splendid than a university.

In these days of broken frontiers and collapsing values, when the dams are down and the floods are making misery, when every future looks somewhat grim and every ancient foothold has become something of a quagmire, where a University stands, it stands and shines; wherever it exists, the free minds of men, urged on to full and fair enquiry, may still bring wisdom into human affairs.

There are few earthly things more beautiful than a University.

It is a place where those who hate ignorance may strive to know, where those who perceive truth may strive to make others see; where seekers and learners alike, banded together in the search for knowledge, will honour thought in all its finer ways, will welcome thinkers in distress or in exile, will uphold ever the dignity of thought and learning and will exact standards in these things.

They give to the young in their impressionable years, the bond of a lofty purpose shared, of a great corporate life whose links will not be loosed until they die.

They give young people that close companionship for which youth longs, and that chance of the endless discussion of themes which are endless, without which youth would seem a waste of time.

There are few things more enduring than a University.

Religions may split into sect or heresy; dynasties may perish or be supplanted, but for century after century the University will continue, and the stream of life will pass through it, and the thinker and the seeker will be bound together in the undying cause of bringing thought into the world.

Masefield's eulogy was appreciated by his audience, but this inspired, idealistic outlook has gradually become diluted by materialism. During the last twenty years, more and more students have looked upon a university education merely as a convenient stepping-stone to a comfortable life with a good income, as I have discovered when discussing careers at graduation time. Of course there have always been some students with this attitude—as is documented by the distich about learning and scholarship (*Wissenschaft*) (2) published jointly by Goethe and Schiller in 1797.

> Einem ist sie die hohe, die himmlische Göttin, dem anderen
> Eine tüchtige Kuh, die ihm mit Butter versorgt.
>
> [Some regard her as a lofty, heavenly goddess, but others,
> Merely as a useful cow, furnishing plenty of butter.]

Another disturbing attitude has crept into university life in recent years: a cynicism about basic research. Because knowledge has at times been misused, students question the virtue of the search for new knowledge. They believe some of the old academic ideals to be irrelevant, or even wrong, in the face of the urgent practical problems which confront the world. Some say that they are disillusioned by the shortcomings of academics who are self-seeking rather than idealistic. But such students are in a minority, and idealists of the old type are still about.

Throughout my undergraduate years, lecture halls and laboratories were very overcrowded because German universities had no control over the intake of students. Everybody who had passed the school-leaving examination of a *Gymnasium* was entitled to be admitted by the university of his or her choice. At the time I went to Göttingen, the newly demobilized older students added to the congestion. The only restraint on the number of students—and it was a significant one—was the size of the parental pocket. Students had to depend almost entirely on parental support. Scholarships were virtually non-existent, and because of the high unemployment, casual or vacation jobs were few; in any case, the intensity of the university courses allowed no time for money-making activities.

At Göttingen, the lectures in anatomy, chemistry (by Adolf Windaus), and physics (by Robert Pohl) were first-rate. At weekends I took part, as a hobby, in botanical excursions with the Professor of Botany, Albert Peter.

I stayed at Göttingen only until the end of the summer of 1919. I had originally chosen to go there because it was fairly near home (sixty miles), at a time when railway services were unreliable. During the summer of 1919, travelling conditions improved and in the autumn I transferred to Freiburg im Breisgau, making use of the freedom accorded to students to move from one university to another at will.

Freiburg University had a high academic reputation. Another attraction for me was that it lay in beautiful surroundings at the foot of the Black Forest, an ideal spot for walking, skiing, and cycling.

Soon after settling in, I received a message that my mother (then in her forty-ninth year) had died suddenly during an influenza epidemic. This was of course a great shock; I felt I had lost an irreplaceable support.

During the lectures at Freiburg, professors sometimes referred to their own research, and what they said aroused in me a keen desire to undertake research myself. So, in the summer of 1920, I asked the Professor of Anatomy, Eugen Fischer, whether I could join in research in his department. He referred me to his second-in-command, Wilhelm von Möllendorf, who accepted me to participate in his work on the theory of histological staining. His aim was to explain selective staining of tissue structures on the basis of physico-chemical principles. He hoped that a better understanding of histological staining would improve the existing trial-and-error methods of finding good staining techniques and perhaps also provide information on the nature of the structures which react with the dye. The work I did under his guidance led to my first publication [1]. The experience I gained while I was there had a significant effect upon my outlook, in that it brought home to me the importance of chemistry and physical chemistry to biology. It prompted me to do a good deal of reading in these subjects and it thus proved to be the germ of my later biochemical studies.

Another preclinical teacher who made a lasting impression on me was Franz Knoop, Professor of Physiological Chemistry. He gave me my first glimpse into the field of intermediary metabolism, through his discussion of the metabolism of fatty acids and the pathway of β-oxidation which he had discovered. Knoop liked to mention that although he had established himself internationally as an outstanding investigator by the time he was twenty-nine, he did not obtain a paid position until he was thirty-seven. He intended to illustrate the then common attitude that fundamental research was not an activity for which people were paid. It was regarded as a privilege of those who were financially independent—as little more than a hobby. (If the research productivity of the universities was substantial, it was largely because unpaid doctoral students provided the necessary extra pairs of hands.) University staff had to earn their living by teaching, and those in clinical subjects by looking after patients. Any research had to be carried out in 'spare time'. Until the Second World War this was the prevailing attitude; only occasionally were a few lucky people paid to do research. I was one of these fortunate few from 1926 until 1930 when I was working for Otto Warburg.

Among the friends I made at Freiburg was a fellow student, Hermann Blaschko, with whom I shared many interests, particularly in medical research. We have remained in close contact to this day. In the course of our careers there have been long periods when we worked in close proximity—in Berlin, Cambridge, and Oxford.

I passed the *Physikum* (equivalent to the Second M.B. of most British medical schools) in the summer of 1921. I stayed at Freiburg for one more semester (six-month term) to begin clinical studies, including pathology, under Ludwig Aschoff, a most inspiring teacher. I moved to Munich at the start of the winter semester and completed my clinical course there (with the exception of one semester at Berlin during the winter of 1922–3). I was attracted by the excellence of the Munich Clinical School, which had many famous people on its staff, Friedrich von Müller, Ferdinand Sauerbruch, and Emil Kraepelin among them. I was also lured to Munich by the high quality of its theatres, opera, museums, and its beautiful countryside—the Bavarian Alps.

It was during my time at Munich that the runaway inflation of the German currency and the concomitant economic depression reached their peak (3, 4). The inflation had begun during the First World War and had steadily accelerated, as the table (p. 21) shows.

In 1920 and 1921 the value of the mark dropped significantly from month to month. In 1922, after the assassination in June of the German Foreign Secretary, Walther Rathenau, the downward trend gained dramatic momentum because foreign confidence in Germany's economic recovery dwindled. Week by week, then day by day, and eventually, in November 1923, hour by hour, the mark dropped and dropped.

As I have already mentioned, most of the students of my generation, myself included, were entirely dependent upon parental support, but parental earnings rarely kept up with inflation. This applied, for example, to all the professions and the civil service. The living habits of most students had therefore to be very simple, with spending restricted to essentials, though until the early part of 1923 conditions were bearable. All

German inflation after the First World War

Date	Value of US dollar in German marks
July 1911	4.2
July 1919	14
July 1920	40
July 1921	77
July 1922*	493
January 1923†	18 000
July 1923	353 000
September 1923	100 000 000
13 November 1923	840 000 000 000
14 November 1923	1 260 000 000 000
15 November 1923	2 520 000 000 000
20 November 1923	4 200 000 000 000

*Rapid fall after assassination of Walther Rathenau, German Foreign Secretary

†Further acceleration of downward trend after occupation of Ruhr zone by French and Belgian troops as sanction for Germany's failure to deliver reparations.

government-controlled services were relatively cheap because the authorities were slow to adjust prices to the changing values. This applied to the transport and postal services and also to the theatres, concert halls, and museums. I learned later (5) that a British visitor had been able to buy the best seat in a first-rate opera house for the equivalent of sixpence—the cost in Britain at that time of sending four inland letters. Foreign visitors flooded into Germany to take advantage of the situation. Shops and restaurants were much quicker to increase their prices. At the height of the inflation, some would not put a price on their goods or meals until the rate of exchange against the dollar had been announced for the day.

My own small savings vanished. When my mother died I had inherited about 2000 marks—at that time worth about £50. This I had kept in a savings account but in November 1922 I decided to spend it before it became entirely worthless. It was still enough for just one book which, together with its receipt dated 11 November, is still in my library. It was *Die Philosophie des Als Ob*, by Hans Vaihinger (6), whose philosophy (sometimes called 'fictionalism') was much discussed in Germany at that time. I found it very interesting, but the long-term impact of the book appears to have been slight.

Although the students were living under primitive conditions, they retained their zest for both study and outside pursuits. My interest in the scientific aspects of medicine was deepening and gradually it became clear to me that I would not, after all, follow my father into otolaryngology, as he had rather hoped.

I passed my final examinations in Munich in December 1923 with first-class marks, but in order to qualify for a licence to practise it was necessary for me, as for all aspiring doctors, to work for a year in a recognized hospital. This caused problems, because it was difficult to find a place in a good hospital even if one were prepared to work for nothing. While I had been studying in Berlin in the winter of 1922-3 I had made up my mind to spend this hospital year there in a department of *Innere Medizin* (general medicine). Berlin was at that time an outstanding centre of excellence, not only in every academic field, but also in the world of theatre, music, and other branches of entertainment. In brief, it was a great cultural centre (7).

I wanted to do general medicine, because it seemed to offer the best scope for developing my research interests, and I applied to several of the best municipal hospitals connected with the university, but without success. Eventually I secured an unpaid post at the Third Medical Clinic, which was under the direction of Professor Alfred Goldscheider.

Goldscheider, then near retiring age, was a good teacher. He was also a good clinician with a distinguished research record. As a young man he had made the fundamental discovery that the sensations of heat, cold, touch, and pain are perceived at different points on the skin. But he was not much in evidence or readily accessible, though he gave regular lectures. Like the majority of senior medical academic staff, he also engaged in private practice—a lucrative but time-consuming pursuit.

Many people at the hospital took part in its research activities and were encouraged to do so, but the quality of the research was mixed. Much of it was carried out for reasons of prestige rather than to satisfy genuine scientific interest and the clinical research workers often lacked proper training in scientific methodology. Nevertheless, for me the research atmosphere was favourable. Among my colleagues there were a number of young, enthusiastic people, all most anxious to talk about their work. Foremost among them were David Nachmansohn and Bruno Mendel, who became my life-long friends and distinguished themselves as researchers. Both had to leave Germany when Hitler came to power in 1933. Bruno Mendel went first to Holland and continued his work on cancer cells, but anticipating the invasion of Western Europe, he accepted in 1936 an invitation to join the Banting Institute in Toronto. After the war ended he was invited to the Chair of Pharmacology at Amsterdam and in 1957 he was elected to Fellowship of the Royal Society. David Nachmansohn went first to Paris, then to Yale, and finally to Columbia University in New York where he became a world leader in the field of neuro-chemistry.

My clinical work at this time was in a section headed by Dr Anneliese Wittgenstein, who was particularly interested in the treatment of tabes dorsalis. This disease, which was then very prevalent, is a late manifestation of a syphilitic infection of the nervous system, causing severe pain, paralysis, and blindness. It was very resistant to anti-syphilitic drugs, and this was thought to be due to the so-called 'blood–brain barrier' which prevents drugs from reaching the nervous tissue. Dr Wittgenstein was anxious to do some research in this field, but was uncertain where to start or how to go about it.

I suggested that we might carry out an investigation based on the assumption that physical–chemical factors (the Donnan equilibria) play some role in the distribution of substances between the blood, tissue, and cerebrospinal fluid. Using intravenous injection of synthetic dyes, whose distribution in the body is easy to trace, we established that only the anionic but not the cationic dyes reached the cerebrospinal fluid. This finding we hoped would be relevant to the penetration of drugs into the nervous system. The work [6, 7, 8, 9] was frequently quoted in subsequent years but I see no reason to be proud of it. My only excuse for its mediocrity is that I had no guidance of any kind in the design of experiments, in experimental techniques, the critical analysis of data, or the writing of papers.

Shortly after I left the Clinic in Berlin, I witnessed for the first time one of those academic intrigues which were not uncommon. It concerned the formal admission of Anneliese Wittgenstein as a recognized university teacher, an important milestone in the career of an academic. Undue pressures in favour of Anneliese had been brought to bear by a member of the Faculty Board on Professor Goldscheider, so that he decided that she should have preference over Bruno Mendel. To those who were in a position to make a fair comparison, this decision seemed grossly unjust. Anneliese was a reasonably competent clinician, but as an academic—as a teacher or researcher—she was insignificant. The thesis which she submitted to the promotion board was based entirely on the work she and I had done together and whose theoretical background she had been unable to understand. Bruno Mendel, on the other hand, as his subsequent career proved, was one of the most original and promising members of the Clinic. He had, moreover, already published a number of major papers. Albert Einstein, who was on friendly terms with Bruno, heard of the affair. He made independent checks and, characteristically, worried about it sufficiently to write to Goldscheider pointing out the error of judgment. Goldscheider was upset and tried to justify the decision in a lame and apologetic letter. To this Einstein replied simply, *'Dixi et animam salvavi meam'* (8) (I have spoken and saved my soul), a passage of biblical origin well known in Germany and not infrequently quoted.

My twelve months at the Third Medical Clinic convinced me of the importance of chemistry to research in medicine and also of the inadequacy of my own chemical knowledge for the kind of research I was visualizing. As the likelihood of making a living as a 'pure' research scientist seemed faint, I thought I would do better to aim at becoming a practising clinician and try to combine some research with the clinical work.

The realization that research in clinical medicine could benefit greatly from biochemistry was generally gaining ground, and the Department of Chemistry at the Pathological Institute of the Charité Hospital, under Professor Peter Rona, provided special informal courses to train medical graduates in chemistry and biochemistry. I decided to join this Department (unpaid, of course) and started to work there early in 1925. My father agreed to continue to support me financially, though rather reluctantly because he questioned whether research, even when applied to clinical medicine, offered much prospect of earning a reasonable living. Another reason for his hesitation was that the inflation had made his financial position rather shaky.

The good training facilities in Rona's laboratory attracted a number of keen young people who wanted to prepare themselves for research involving biochemical methods and concepts. During my stay these included David Nachmansohn, Karl Meyer, and Robert Ammon. Hans H. Weber was a staff member. Fritz Lipmann and Ernst Chain also worked for a period in this laboratory, though not at the same time as myself.

I left Professor Rona in December 1925, when a lucky chance led to my obtaining my first paid post as Research Assistant with Otto Warburg at the Kaiser Wilhelm Institut für Biologie at Berlin-Dahlem. My good luck hinged on friendship: my own friendship with Mendel, Mendel's friendship with Einstein, and Einstein's friendship with Warburg. In 1925, Bruno Mendel and his wife Hertha were invited to a small dinner party at the Einsteins'. One of the other guests was Otto Warburg, who had been known to Einstein, through his father, since about 1910. The table order placed Warburg next to Hertha, who told him that her husband was keenly interested in Warburg's cancer work. This started a scientific friendship between Mendel and Warburg. Soon they were telephoning each other every evening to discuss their experimental results and ideas. During one of these discussions Warburg said that he had too few collaborators to carry out all the experiments needed to put his ideas and theories to the test. Mendel had a high opinion of my potential as a scientist and suggested to Warburg that I would be a suitable research assistant for him. Warburg replied that it would be a matter of obtaining funds to pay me, whereupon Mendel approached a potential benefactor. Mendel was a very impressive man, and very persuasive. He succeeded in raising some money and was able to offer Warburg enough for my first year's salary. This was just adequate for a bachelor, and by being careful I managed to have something left for travelling during the summer vacation.

That chance should play a major part in the progress of one's career was by no means unusual. Indeed, getting a junior position very much depended on knowing people, on having a personal connection, on having an effective supporter. This was because it was not customary in Germany to advertise academic positions—in contrast to the British rule that all posts must be advertised so that anybody who feels qualified can make sure that his name is known to the selectors. In Germany, the annual meetings of learned and professional societies were used to establish contacts between those looking for a post and those looking for new staff.

At the interview which led to my appointment Warburg warned me that a post in his laboratory was not likely to help me towards a university career. German professors, he said, regarded him as an outsider; he was not on good terms with many of them, and he could not help me to get even a junior university post when my temporary appointment with him came to an end. I could learn a lot in his laboratory, he said, but if it were my aim to make a university career, 'Then you had better attach yourself to some old ass of a professor'. Later he confessed to the belief that if one wanted to have a successful academic career in Germany, it was important to ingratiate oneself with an influential professor. His words at our first interview did not deter me; his reputation as one of the leading German biochemists was already firmly established and I was not too concerned about my long-term future.

Warburg kept his word on both counts: he taught me a great deal, and he did not actively help me to find a post when he asked me to leave his laboratory some four years later.

3
IN THE LABORATORY OF OTTO WARBURG
(1926–1930)

Otto Warburg was the most remarkable person I have ever been closely associated with. Remarkable as a scientific genius of the highest calibre, as a highly independent, penetrating thinker, as an eccentric who shaped his life with determination and without fear, according to his own ideas and ideals. I have tried to do justice to his personality and achievements in a biography [348].

Among all my teachers he had by far the greatest influence on my development and I owe him an immense debt. He set an example of high standards in research and in general conduct, particularly in his genuine dedication to his chosen area of activity, the advancement of scientific knowledge. His dedication manifested itself in his long and regular working hours and his contempt for those who tried to further their careers by jockeying for position, by hobnobbing with and courting the influential, or by publishing trivia for publishing's sake. He was prepared to take infinite pains with every aspect of his work and prided himself that his experimental findings and his presentation and interpretation of them, were clear-cut, reliable, and able to stand the test of time. He also took pride in the fact that when he found that he had made a mistake (which did not happen often) he would admit it and publish a correction without delay.

In his work, Warburg followed three main lines of investigation: photosynthesis, cancer, and the chemical nature of the enzymes responsible for biological energy transformations, i.e. of oxidations and reductions. These three subjects (in many ways interlinked) occupied him throughout his life from 1919 to 1970. He was convinced that major scientific pioneering depended very much on the development of new research tools, and his innovations opened up large areas for research. His new manometry (the measurement of gas pressure) greatly simplified the investigation of reactions involving gases, such as the uptake of oxygen and formation of carbon dioxide in living cells or the reverse process in photosynthesis in plants.

Manometry was gradually replaced by spectrophotometry, and Warburg was instrumental in this development also; long before spectrophotometers became commercially available in the later 1940s Warburg had been building them in the laboratory, beginning in the 1920s. He invented the 'tissue slice technique' which made it possible to make accurate measurements of metabolic processes in pieces of isolated tissue whose structure was essentially intact. He developed new methods for the purification and crystallization of enzymes. He made decisive contributions to the identification of the chemical nature of the catalysts of cell respiration and he explained the manner of their action; in 1931 he was awarded the Nobel Prize for his discovery of the catalytic role of iron porphyrins in biological oxidations. It was Warburg who identified nicotinamide as the active group of enzymes, who made the first quantitative measurements of lactic acid production in many animal tissues and who discovered the high rate of lactic acid production in cancer tissue. He introduced unicellular algae as experimental material for the study of photosynthesis and measured the efficiency of photosynthesis.

Warburg made a deep impression on all who came into contact with him, inside and outside science, and irrespective of whether or not they agreed with his views. Whenever people who knew him were together, their conversation turned sooner or later to the subject of his personality. What fascinated people, apart from his penetrating intelligence which gave him a deep understanding of many aspects of his science and of affairs generally, was his intellectual honesty and straightforwardness, his singularity of purpose and his industry, the wit and humour of his pertinent comments on affairs—scientific and other, his generosity in helping people in the laboratory, and his eccentricities. Those who knew him well were very ready to overlook weaknesses, his touchiness, his resentfulness, his prejudices, and his harshness against those whom he regarded unjustifiably as his 'enemies'. Thus David Keilin, whom Warburg had at times abused and ridiculed in polemic articles (1), was keen to support Warburg's nomination to the foreign membership of the Royal Society.

Warburg was a man who fully devoted his gifts to the advancement of science. He continued to work in the laboratory until eight days before he died in his eighty-seventh year from a pulmonary embolism. It had been his view that as long as he could control the work in the laboratory he should carry on; he would retire as soon as he felt he could no longer check every detail personally. To Warburg, that time never came.

When I joined Warburg's laboratory I knew very little about his personality. I had read some of his work and I appreciated that he had the reputation of being an outstanding scientist. Bruno Mendel had told me that he was exceptionally stimulating and profound, and that he could be charming. My expectations were great; indeed, I was elated, not only because I had my first paid job but also because it was the best job I could have dreamed of—the opportunity of engaging full-time on research at the Kaiser Wilhelm Institute for Biology under the guidance of one of the greatest, perhaps the greatest, scientist in my field of interest in Germany.

The Institute for Biology was one of the Institutes of the Kaiser Wilhelm Gesellschaft (in 1948 renamed the Max Planck Gesellschaft). The Society had been set up in 1910 to provide outstanding scientists with the best of research facilities. The attitude of its founders was clearly expressed by Emil Fischer when he successfully persuaded Richard Willstätter to abandon his professorship at Zurich and join the Society with the words, 'You will be completely independent. No-one will ever interfere. You may walk in the woods for a few years if you like; you may ponder over something beautiful' (2).

This policy of the Kaiser Wilhelm Gesellschaft, based as it was on the utmost care and ability in selecting the right people, did, on the whole, pay magnificent dividends: apart from Warburg and others whom I have already mentioned, Max von Laue, Fritz Haber, Otto Hahn, Lise Meitner, Michael Polanyi, Carl Neuberg, and many others made the fullest use of the opportunities it offered.

Berlin-Dahlem was the main campus of the Kaiser Wilhelm Gesellschaft during the 1920s. It had been built in spacious grounds in a pleasant, quiet, green suburb some six miles from the city centre. Several buildings housed a number of departments, each designated by the name of its head rather than by a subject of research—another reflection of the freedom afforded to the leading scientists in their choice of problems. Contact between the various departments was very close. By the late 1930s, within fifteen years of its foundation and despite the upset caused by the First World War, Dahlem had become a world centre of excellence in scientific research. Not only did it attract many of the best scientists in Germany but also brilliant people from all over the world. Today most of the buildings are part of the Free University of Berlin.

When I started work at the Kaiser Wilhelm Institut für Biologie there were, in addition to Warburg, five research workers: Erwin Negelein, Franz Wind, Hans Gaffron, Robert Emerson, and Fritz Kubowitz. There was also a *Diener* (an odd-job man). Other researchers who joined the laboratory whilst I was there were Walter Christian, Walter Kempner, Akiji Fujita, Erwin Haas, Werner Cremer, and Joseph Donegan.

The outstanding work done in Warburg's laboratory just before and during my stay included:

the discovery of the aerobic glycolysis of tumours;
the discovery of the general occurrence of the Pasteur effect;
the first accurate quantitative measurements of cell respiration and cell glycolysis;
the discovery of the carbon monoxide inhibition of cell respiration and its light-sensitivity;
the identification of the oxygen-transferring enzyme in respiration (now referred to as cytochrome A_3) as an iron porphyrin;
the development of spectrophotometric methods of analysis (incorporated twenty years later by Beckman into his commercial spectrophotometer, the 'black box');
the discovery of copper in blood serum and the fall of its concentration is anaemias.

All the work was done in one large room equipped with chemical benches. There were also three instrument rooms and in 1928 another chemical laboratory was added which housed four people.

LABORATORY OF OTTO WARBURG (1926–1930)

In addition to Warburg's team, the building accommodated five other departments headed by Carl Erich Correns, Richard Goldschmidt, Max Hartmann, Otto Mangold, and Otto Meyerhof, a former student of Warburg and already a Nobel Laureate. Meyerhof and his team were of special importance to us in Warburg's laboratory because they were engaged on very similar work. At various times the Meyerhof team included Karl Meyer, Karl Lohmann, Fritz Lipmann, Hermann Blaschko, David Nachmansohn, Severo Ochoa, Dean Burk, Ralph Gerard, Frank Schmitt, Louis Genevois, and Ken Iwasaki. Also working in the same building were Viktor Hamburger, Curt Stern, and Joachim Hämmerling, three young biologists who were to leave their mark on later scientific developments.

Meyerhof's laboratory made decisive contributions to what is now called the Embden–Meyerhof pathway of glycolysis. It laid the groundwork which led to the discoveries of hexokinase, aldolase, and other enzymes. Momentous discoveries were made by Lohmann—ATP (first identified as a cofactor of glycolysis), and the 'Lohmann reaction' (the interaction between ATP and creatine).

A major factor behind the outstanding achievements in the two laboratories was the inventiveness of their leaders: Warburg originated the tissue slice technique, manometry, and spectrophotometry; Lohmann found the way to distinguish the many phosphate esters by measuring their rate of hydrolysis at 100 °C in 2N hydrochloric acid.

A remarkable feature of both teams was their small size. Altogether no more than twenty or thirty people participated in these great developments. Meyerhof had only four or five small rooms and the total number of his collaborators at any one time never exceeded five. He had only one trained technician, Walter Schulz, and a part-time typist. The technician was Meyerhof's personal assistant who, together with a *Diener*, looked after the laboratory housekeeping and maintained the apparatus. Warburg had no technicians in the usual sense of the term. His long-standing collaborators (Negelein, Kubowitz, Christian, and Haas) were, in fact, research assistants. They had been trained as instrument mechanics in the workshops of the Siemens factories in Berlin and had learned how to handle instruments and how to make accurate measurements; Warburg taught them all the chemistry they needed.

Not only did the genius of leadership shown by Warburg and Meyerhof make an outstanding contribution to biochemistry and inspire their deeply motivated and committed young collaborators: in the process the two men educated the future generation of leading scientists. This happened naturally; I believe something of this sort always will happen naturally. Born leaders attract born followers who develop into leaders—as long as bureaucracy and erroneous concepts of equality do not interfere. Only the perceptive student will derive the full advantage from this sort of training, whilst the teacher who spends a large proportion of his time away from his science—in committee rooms, on travels, and on administrative matters—easily sets the wrong kind of example by indicating that he considers such activities as more important than doing experiments.

Although we all worked long, hard hours in the laboratory—from eight in the morning until six at night, six days a week—the atmosphere was relaxed. During the brief mid-day interval the younger people from the different departments (in particular those working with Warburg and Meyerhof) would meet in a common-room for a simple snack of eggs, sandwiches, and milk. Coffee- and tea-breaks were unknown. Most of our scientific reading and writing was done at home in the evening, on Sundays, and during the long summer vacation when the laboratory was closed. Warburg liked to point out that the working hours were much shorter than they had been in his younger days. When he had worked in Heidelberg in Krehl's Department of Medicine, for instance, Krehl had often made a round of the laboratories on Sunday evenings and had expected most of the workers to be there. In Meyerhof's laboratory working hours were less rigid but hardly shorter than ours.

Within his laboratory, in the day-to-day contacts with his collaborators, Warburg was always friendly and helpful, as long as they treated him with due respect. We never questioned Warburg's autocratic control; in our eyes it was justified because of his outstanding intellect, his achievements, and his integrity, qualities which we admired enormously. On the whole he was a benevolent dictator but he could also be fierce.

He once dismissed a research worker on the spot, when he thought the man had not shown him proper respect and courtesy. It seemed that to Warburg absolute control was essential to maintain high standards in work as well as in personal conduct. It was autocratic rule at its best, however. He never exploited the junior, as does autocratic rule at its worst. Democratic rule at best makes full use of pooled resources, but at worst it creates a situation where ignorance and obstruction can prevail over competency and efficiency.

Money was very limited. Several people in the laboratory received no salary or grant. When Hermann Blaschko had asked Meyerhof if he might work in his laboratory, Meyerhof agreed, but told him, 'I cannot give you any payment', to which Blaschko replied, 'I did not expect any'. Fritz Lipmann was not paid during his first two years with Meyerhof; Severo Ochoa received no pay at all. I was lucky to receive a starting salary of 300 marks per month, roughly £180 a year. After a year Warburg raised this to 400 marks per month (£240 a year), I believe from his own resources. It is difficult to equate this to present values, but in order to manage, I had to live frugally and count every penny. If I was careful I could afford one modest holiday a year, and to go to the theatres and concerts occasionally if I went into the cheapest seats. There were no travel grants for attending scientific meetings, but this did not cut us off from meeting other scientists or learning about new developments. There were plenty of such opportunities in Berlin itself: the colloquia at Dahlem and the Berlin Chemical and Medical Societies.

Who, then, financed young researchers? As far as I know, it was our parents, many of whom could ill afford it. Inflation, as I have already mentioned, had devalued the pre-war mark by a factor of 10^{12} (a million million) by the end of 1923. Nevertheless, parents were willing to make sacrifices in order that their sons should have a sound training. Naturally, we felt very uncomfortable, knowing ourselves still to be a financial burden at the age of twenty-five or thirty.

Most of us were not married; we could not afford it. When I was twenty-six I wrote to my father saying that I would like him to meet a girl with whom I had a growing friendship. He sent a very brief and rather unusual reply. He merely referred me to

Rocco's 'Love will not suffice' aria from *Fidelio*; I looked it up and found:

> Wenn sich nichts mit nichts verbindet
> Ist und bleibt die Summe klein.
> Wenn man bei Tisch nur Liebe findet,
> Wird man nach Tisch hungrig sein.
>
> [When a farthing marries farthing
> Small's their sum, and small remains.
> If naught but love is on the table
> You will suffer hunger pains.]

I took this to heart.

It was understood in those days that an academic career meant accepting a very modest standard of living. My fellow workers and I were motivated by a keen dedication to our work and by the hope that when we had received a thorough training (which we expected to complete by the time we were about thirty) we would find work which would be professionally satisfying and provide a reasonable income. The sacrifices imposed by a long period of training meant that only the keenest stayed the course. Most of us were medical graduates and could have branched off at any time into a relatively lucrative medical career, but we preferred research.

The conditions at Dahlem were, of course, not unique. Erwin Chargaff, a contemporary who was at that time working in Vienna wrote (3), 'No-one who has entered science within the past 30 years or so can imagine how small the scientific establishment then was. The selection process operated mainly through a form of an initial vow of poverty. Apart from industrial employment, important for a few scientific disciplines such as chemistry, there were few university posts, and they were mostly ill-paid.'

Nowadays there are many students who insist upon their 'right' to be fully supported by the state. We neither expected nor felt we had any rights, even at the post-graduate and post-doctoral levels. We were satisfied if we could work and learn under reasonable conditions, and we worked hard. We did not feel entitled to expect, let alone demand, anything, before we had learned a lot and were able to make at least a small contribution to society.[5]

Despite our slender means we were on the whole very happy. We felt that we were receiving a first-rate training and were doing something worth while. We had no undue worries about our long-term futures although these looked very uncertain in view of the economic and emerging political difficulties in Germany. I do not think that any of us assumed or anticipated a particularly successful career. Being close to some of the giants of science we felt very small, and Warburg did little to boost our self-confidence.

The relationship between Warburg and Meyerhof on the one hand, and the official representatives of German physiology and biochemistry (the German university departments) on the other, were somewhat strained. As I have already mentioned, Warburg regarded himself as an outsider; Meyerhof, too, although perhaps less so. It is remarkable that the German universities did not properly acknowledge Meyerhof's qualities. When he received a Nobel Prize in 1923, at the age of thirty-nine, he held the relatively junior position of 'Assistant' in the Physiological Institute of Kiel University. He had been passed over for a junior professorship (Professor Extraordinary) in favour of a man called Pütter, whose claim to distinction was, and remained, slight. It was only after the Nobel Prize that Warburg succeeded in persuading the Kaiser Wilhelm Gesellschaft to offer Meyerhof an appropriate post which would give him, for the first time, reasonably good research facilities.

To try to understand why these two great men had a sense of being outsiders, one must first appreciate the Cinderella treatment of biochemistry by the German universities at that time. There were very few chairs or departments of biochemistry or physiological chemistry. Independent departments existed in four universities only: at Frankfurt, with Embden at its head, and at Freiburg, Tübingen, and Leipzig. In the other universities biochemistry was a sub-section of physiology and these section heads did not have the rank of full professor. In some, the professor of physiology was essentially a biochemist. This applied to Heidelberg, for instance, where Kossel, the Professor of Physiology, did first-rate work on protein chemistry. Because of this situation Leonor Michaelis, one of the brightest biochemists of his time, could not be absorbed into the German university system and was earning a living as a clinical biochemist in one of Berlin's municipal hospitals. He left Germany in 1921 for Japan and moved later to the Johns Hopkins University and the Rockefeller Institute.

We felt much encouraged and grateful, therefore, when F. G. Hopkins, in his presidential address (4) at the International Physiological Congress held in Stockholm in 1926, commented on the neglect of biochemistry in German universities and spoke about the importance of 'specialised institutes of general biochemistry'. He referred to the appeal by Hoppe-Seyler forty-nine years earlier in Volume 1 of his *Zeitschrift für Physiologische Chemie* in 1877, that institutes of biochemistry should be set up in the universities. (In 1877 the only such institute was in Strasbourg with Hoppe-Seyler at its head.) His appeal was immediately opposed by the physiologist E. Pflüger in his journal on the grounds that physiology should not be fragmented. Hopkins remarked that Hoppe-Seyler's plea for the recognition of biochemistry as an independent subject had not yet, nearly half a century later, found a proper response in his own country. He added, 'It is difficult to see how Germany can continue to lead along the path which for a long time she has trod almost alone.' He emphasized that, in general, the academic centres in Europe were behind those of America in this respect. His remarks were, incidentally, made at the suggestion of Franz Knoop for the benefit of the German readership and they were reprinted in translation in the Münchener Medizinische Wochenschrift (5).

My research in Warburg's laboratory was controlled by Warburg's own interests. He wanted every member of the laboratory to be part of his personal team and allowed scope for initiative and independence only if it fitted within the framework of his own field. His main interest at that time was the metabolism of tumour tissue and the nature of the catalysts responsible for cell respiration. During my association with him I participated in a minor way in the work on the 'action spectrum' of the 'respiratory enzyme' (now called cytochrome oxidase) for which he was awarded the Nobel Prize in 1931. He asked me to work out a simple system, a model, by which oxidations by molecular oxygen (such as cell respiration) are inhibited by carbon monoxide, and where this inhibition is sensitive to light. I found that under certain conditions the oxidation of cysteine catalysed by haem, an iron porphyrin derived from haemoglobin, meets this requirement [17]. This made it possible for Warburg to check on a simple model the theory he had developed for the complex system of intact yeast cells.

Other joint work with Warburg led to the demonstration that traces of copper are present in blood serum [15, 16]; that the copper concentration is constant under normal conditions but varies characteristically under pathological conditions, in anaemia for example. This work arose from a discovery by Warburg that, in pyrophosphate buffer, copper ions are the only heavy metal ions that catalyse the oxidation of cysteine. Warburg exploited this phenomenon for the elaboration of a method for the determination of copper which proved to be far more sensitive and far more specific than any other method available at that time, i.e. to an accuracy of less than one microgramme. It is now known that the copper ions are loosely attached to a protein called ceruloplasmin.

In all, I published sixteen papers during my four years in Warburg's laboratory.

During the two month summer vacation I tried to combine recreation with learning languages—as well as my scientific writing and reading, of course. At school I had learned the rudiments of English and French, but no more, so in 1926 I spent a few weeks in the French Alps (Argentières, Talloires) and in Paris. In 1927 I visited Belgium. These countries were my first choice because it was very much cheaper to go there than to England. By 1928 I had saved enough to cross the Channel and I booked myself in at a modest boarding house at Sandown on the Isle of Wight. I travelled by sea from Hamburg to Southampton, stayed there overnight and arrived in Sandown just before lunch. At an adjacent table in the dining-room sat a middle-aged couple, a Mr and Mrs Sayer from Wembley, Middlesex, with a daughter in her early teens. As soon as I had finished my meal, Mr Sayer offered me a cigarette and started to chat. When I told him that I was German and that I was keen to improve my English during my holiday, he and his family at once started to help me in a very friendly and entirely unobtrusive way. Sayer was a teacher at a technical college and knew how to go about my tuition. We talked, he corrected me, he invited me to join in their excursions, and he introduced me to beach cricket. Later he invited me to visit them at their home in Wembley at the end of my vacation when I would be spending some time in London. Thus, through these kind people I entered a private English home for the first time and had my first experience of British friendliness and human warmth. I had never met anything like it before, extended as it was to a complete stranger, a young nobody, a mere fellow human being. True, I had made friends of contemporaries in Germany and had occasionally been invited to their homes, but this kind of friendliness and helpfulness was not commonly found in Germany nor, as far as I know, elsewhere in central or western Europe.

A year later, when I took advantage of an opportunity to spend two months in the United States, visiting Boston, New York, and Woods Hole, I was to learn just how much this friendliness distinguishes the Anglo-Saxon societies from those of continental Europe. It impressed me enormously, so that in 1933, when I was forced to seek a new home outside Germany, I knew where I wanted to go.

The purpose of my visit to the United States was to attend the Thirteenth International Physiological Congress at Boston. It was made possible by a small grant (arranged by Warburg) from the German Physiological Society to defray part of the travelling expenses. Further financial help came through Robert Emerson of Harvard who initiated an offer from Dr Joseph C. Aub to spend three weeks in his laboratory before the conference started, to teach manometry. (Among my 'students' was Ovid Meyer, who later became Professor of Medicine at the University of Wisconsin.) I shared lodgings with a contemporary post-doctoral student, Marcel Florkin of Liège, whom I was to meet later on many occasions.

The Congress itself was a great experience because it brought me into contact with a number of leading physiologists and biochemists, including the Coris, van Slyke, Michaelis, Anson, Knoop, Szent-Györgyi, Abderhalden, Klothilde Gollwitzer-Meier, and many others. Following the Congress, there was an excursion to New York and while we were there I unfortunately developed a minor abscess on my leg. Although it was not serious, I was told to lie still and the Congress Bureau arranged a bed for me at the Harkness Pavilion of the Presbyterian Hospital, where I was looked after extremely well. I had several visitors, including Tim Anson and Robert Loeb. Loeb I had not met before, but have been in touch with ever since. His reason for visiting me was because his father, Jacques, had been a friend and admirer of Warburg.

The abscess subsided after a few days and I accepted an invitation from Leonor Michaelis and his wife to spend a week at their summer home at Woods Hole. It proved to be a most enjoyable visit. Michaelis impressed me by his modesty, understanding, and friendliness and by his profound and wide intellect. Through him I met many other people at Woods Hole with whom I remained in contact. Much of the friendliness extended to me stemmed from the fact that I had been associated with Otto Warburg, whose reputation was international.

Whilst I was in Boston, Detlev Bronk got in touch with me. He asked me to visit him in Philadelphia to discuss a possible post for me at a new research institute, The Johnson Foundation, which he was setting up. Bronk had asked Warburg if he could suggest some good staff and Warburg had put my name forward. I met Bronk in Philadelphia but did not feel inclined to pursue the matter. The post was only a temporary one and the research would have been biophysics rather than biochemistry. Only after I returned to Dahlem did I become aware that Warburg had hoped I would accept Bronk's offer.

4
ALTONA AND FREIBURG
(1930–1933)

During 1929 Warburg reminded me from time to time that I could not stay indefinitely in his laboratory, though at first he did not specify a time limit. He stressed that he did not want to keep senior scientists in his laboratory but to fill the few places at his disposal with young research assistants (preferably technicians) and occasional short-term visitors. He seemed rather disappointed that I had not followed up Bronk's suggestion during my visit to the United States, and soon after my return he gave me notice to leave by 31 March 1930.

He offered no help, except as a referee. His unwillingness to do more was, I believe, because he did not think I had sufficient ability for a successful research career. Later, when he had evidence of my capacity for independent research and had witnessed my professional successes, he became much more friendly and helpful; during the last ten years of his life there were moments when he showed me real personal warmth. Charles Huggins told me that Warburg had described me as his favourite pupil; I must say that Warburg never gave me any inkling of such feelings. I was indeed surprised and touched when in August 1970, some three weeks after Warburg's death, I read a dedicatory preface he had written for a volume, *Essays in cell metabolism* (1) presented to me by my associates, past and present, on my seventieth birthday:

When Hans Krebs in 1925 appeared at Berlin-Dahlem in the Kaiser Wilhelm Institute of Biology, he was a modest thoughtful youth, very intelligent and already wise in spite of his youth. At the beginning of his four years' visit you could not see who he was. But at the end you could.

Nearly 40 years later I met him in Dublin at an important meeting. He was now Sir Hans and the chairman of the meeting. Few scientists were more famous. How civilized he is, I thought, if I compared him, even with myself.

I cannot help feeling, disrespectful though it might be, that this favourable opinion was much influenced by my professional success—a Nobel Prize, an Oxford professorship, and a knighthood—and even more by the fact that I had persuaded Oxford University to confer an Honorary Degree upon him, an honour he valued highly.

During my last few months in Warburg's laboratory, I did a great deal of heart-searching about the kind of career I should aim at. When I asked Warburg for advice about going back to medicine or looking for a post in biochemistry or a related subject, his answer was to the point. 'Biochemistry', he said, 'is not a profession, but medicine is. What I mean is this: you might decide to marry and you may then want to buy a house. You may have to go to a banker and ask for a loan. The banker will ask you about your profession and what you earn, and when you tell him that you are a biochemist he won't give you a penny, but if you tell him that you are a doctor you will probably get what you want.' He offered no encouragement whatsoever in the direction of a research appointment—perhaps with some justification: in 1929 biochemistry was certainly not a profession in the sense that one could expect to make a living from it. At that time there were no more than a dozen senior posts for biochemists in Germany. In England there were chairs of biochemistry at Liverpool, Cambridge, and Oxford as well as in a few London schools; attached to each of these was a very limited number of junior posts. Industry, even in biologically orientated fields such as the pharmaceutical and food industries, preferred at that time to employ chemists rather than biochemists. Even ten years later, jobs in biochemistry in Britain were very few and far between. A vivid illustration of this was an advertisement in *The Times* of Monday 20 February 1939:

M.A., Ph.D. (Cambridge) Biochemist, British, 34, offers services at a nominal salary in London in any capacity where he could learn useful occupation.

So Warburg was certainly right in saying that biochemistry was not a recognized profession. Moreover he warned me that widespread anti-Semitism in Germany might thwart any attempt I might make to find a suitable position in a university. I accepted his advice, mostly because I had no reason to believe that I was competent enough to make a successful career in a field where scientific research was an essential part of the work. In the absence of any encouragement from Warburg, and comparing myself to him and the other giants at Berlin-Dahlem, I indeed felt dwarfed. Altogether it was difficult for anybody in my position to assess his own ability and potential. At no stage of a student's career was there any attempt at the kind of careful grading and assessment that is carried out continuously in British universities by allocating marks for performance. Because there were so many students, undergraduates were not as a rule known personally to their teachers. Medical students, in common with others, took only two examinations during their five years of study. Although they were classified into four groups after the examinations, the grading was superficial and nobody took much notice of it. When applying for a post it was not customary to be asked for one's examination grading because it was well known that this often depended upon the whims or prejudices of the examiner. All examinations were oral and taken by one examiner only, so that there was no safeguard for impartial judgement. This system gave students no help in assessing themselves.

Since Warburg did not want to keep me and nobody else offered me a job, I came to the conclusion that my talents were rather mediocre. It was only my keen interest that drove me to keep trying for a position which would give me scope for research. Only one friend of any influence, Professor Klothilde Gollwitzer-Meier, encouraged me not to give up hope of finding a suitable post. She was a much respected clinician and respiratory and circulatory physiologist in Berlin whom I had first met at the Physiological Congress in 1929. After our return from Boston we frequently discussed scientific problems. She believed in my competence and she promised to help me, as she was on good terms with many of the leading German clinicians.

I had decided to aim at a position in a department of medicine at a hospital where research was encouraged, and to begin in a university hospital; this would also give me training in clinical medicine, a standby should I fail in my research ambitions. I wrote to several professors of medicine whose interests lay in the area of metabolic and biochemical aspects of medicine, submitting my *curriculum vitae*. There was no positive response. Warburg let me know that several of them had written to him saying that they might be interested in having a collaborator of his on their staff, but that they needed full information. Warburg told me that he knew from past experience that 'full information' really meant that they wanted to know whether I was Jewish or not.

By March I still had not found a position and my only remaining hope was, that by attending the Annual Congress of Internal Medicine at Wiesbaden in April, I might make a successful personal contact. By tradition, this Congress had become an occasion where senior people looked out for new young collaborators and junior people were available for interview. In the meantime, Professor Gollwitzer-Meier had established two contacts for me. The first was with Professor Leo Lichtwitz at the Municipal Hospital of Altona (a town which has become part of Hamburg). Though not connected with the University, Lichtwitz was one of the best general physicians in Germany and had a special interest in metabolic diseases. He was recognized as an outstanding natural leader and gifted organizer of the care of patients. He published a standard textbook entitled *Clinical chemistry* which was one of the earliest of its kind.

Lichtwitz offered me a post, to begin immediately on 1 May 1930. It was a supernumerary post as Assistant in the Department of Medicine. I would have to spend most of my time in clinical work, but I would have the use of a good research laboratory. Warburg supported my grant application for a few pieces of special equipment.

The second contact which Klothilde Gollwitzer-Meier mediated for me was with Professor Siegfried Thannhauser, at that time in Düsseldorf. He had no immediate vacancy but expressed positive interest. Six months later he was invited to a Chair at the University of Freiburg. Thannhauser asked me to come to Düsseldorf for an interview and subsequently offered me an appointment. By this time I was working at Altona, and receiving an excellent training in clinical medicine. I was resident in the hospital and my clinical duties were heavy, with frequent night calls, but Lichtwitz had provided me with a technician so that it was possible to maintain continuity in my laboratory experiments. The hospital also provided a stimulating environment. Lichtwitz created an excellent intellectual atmosphere; everyone was dedicated to their medical responsibilities and friendly and helpful to each other. David Nachmansohn was again one of my colleagues and another was Arthur Jores, who was keenly interested in clinical and physiological research. I formed lasting friendships with two other new colleagues, Georg Tidow and Walter Auerbach (later Walter Auburn, after his emigration to New Zealand), and also with two of the nursing sisters.

Being a municipal hospital, however, Altona was outside the ordinary academic stream. The position which Thannhauser was offering in Freiburg would take me to the heart of the academic world and the opportunities for research seemed exceptionally good. A new university hospital with excellent facilities was being built and three rooms on the laboratory floor could be allocated to me.

At Altona my research was still influenced by Warburg. As I have said, he had supported my grant application for some equipment, but he had done so on the understanding that I would pursue a problem which he had formulated. This was the measurement of the rates of activity of proteolytic enzymes in malignant tumours. In the early 1920s he had shown that malignant tumours differ from many normal tissues by the fast rate at which they degrade carbohydrate to lactic acid. Some critics had argued that this was not a specific characteristic and that other degradation reactions might also be accelerated in tumours. So Warburg asked me to measure the rate of degradation of protein. It was not an inspiring problem because the answer could be guessed, namely that there were no exceptional proteolytic enzymes in malignant tumours. Moreover the methods of approach were limited. I measured the rate of amino acid formation in broken-up tumour cells after the addition of proteins and indeed found no characteristic differences between tumour cells and normal tissue [22, 23, 24, 26, 27], which was what Warburg wanted to have established. At Altona, too, I had to devote most of my time to clinical work to learn the practice of medicine. My time for research was limited, too limited for a more ambitious research project. I therefore accepted Thannhauser's offer, and took up my post at Freiburg on 1 April 1931.

There were many distinguished people working at Freiburg University in the early 1930s. In the medical faculty there were Ludwig Aschoff (pathology), Wilhelm von Möllendorf (anatomy), and Siegfried Thannhauser (metabolic diseases). Among the scientists were three future Nobel Laureates: Hans Spemann (zoology), George de Hevesy (physical chemistry), and Hermann Staudinger (organic chemistry). Among my other contemporaries who reached distinction were Viktor Hamburger (whom I had known at Dahlem), Rudolf Schönheimer, and Franz Bielschowsky, while one of my co-workers was T. H. (Hannes) Benzinger.

My status at Freiburg was that of 'Assistant' in the Department of Medicine. I had full clinical responsibility for a ward of twenty-two patients and after a few months another, similar, ward was added to my care. During one term I had the additional duty of acting as the Professor's assistant at demonstration lectures, which were given five times a week. At these lectures, patients were presented and their diseases discussed in a broad context. The assistant was responsible for the proper functioning of these demonstrations: he had to see to it that the patient was there at the right time; that the results of all ancillary investigations such as laboratory tests, X-ray films, electrocardiograms, slides, and epidiascopic material were at hand and he also had to operate the projector. All of this took a great deal of organizing and the sum total of my clinical responsibilities was much greater than that of any of my colleagues. The reason for this, which I accepted, was that I had to catch up in gaining clinical experience since I had spent so much of my early post-graduate years in the laboratory. Although my clinical responsibilities consumed much time and energy, and competed with my research interests, I thoroughly enjoyed looking after the patients. It was work which provided immediate day-to-day satisfaction, in contrast to research, which is bound to include many disappointments and frustrated efforts.

The majority of my forty-four patients were not gravely ill and I had plenty of help from junior unpaid assistants and visitors under training. The laboratory facilities were excellent, especially after the Department of Medicine was transferred to an entirely new and modern building late in 1931. Being resident in the hospital and unmarried, I could put in a long working day. Any thoughts of marriage would in any case have been premature because my annual salary was about £150. With the free board and residence I received, this was just enough for a bachelor; for a married man with a home to maintain, it would have been completely inadequate.

At Freiburg I was completely free, for the first time, to follow my own ideas on research and to choose my own subject. I had been allowed to take my own equipment from Altona to Freiburg and was able to augment it by a grant from the Ella Sachs Plotz Foundation of Boston. (The Secretary of the Foundation was Dr Joseph Aub, in whose laboratory I had worked for a few weeks in 1929.) Although the grant was not large, it was enough to make a significant improvement in my experimental resources and to allow me to appoint an untrained junior technician. This appointment was essential because my clinical responsibilities could call me away at any time during the day or night, and a technician could see to the continuity of laboratory experiments—at least during ordinary working hours. I also had help in the laboratory from several medical students and young graduates who wanted to do a piece of research in order to obtain an M.D. degree. My laboratory expenses were eased by a grant from the Rockefeller Foundation to my chief, Professor Thannhauser, who helped and encouraged me in every way. This was my first personal contact with the Rockefeller Foundation, an association which was to continue over three decades and to be of decisive importance to my research. The Foundation's European representative was Dr Robert A. Lambert.

University hospitals were encouraged to include in their research activities programmes based broadly on the fundamental sciences because it was accepted that such activities would eventually be relevant to clinical medicine and the sick. So I decided to base my research on the general biochemical experience I had gained in Warburg's laboratory and to investigate, by way of a new approach, some aspects of protein metabolism. My basic ideas about this new approach were very simple. In Warburg's laboratory I had seen the importance of his 'tissue slice technique' in studying the respiration glycolysis of animal tissues *in vitro* (i.e. under exactly controllable conditions). Until this development, it had not been possible to study respiration glycolysis of animal tissue *in vitro* because all the tissue preparations which had been tried, such as minced or chopped tissue, had lost their metabolic activity. Later developments in mitochondria, in the 1940s, showed why these early efforts had failed to obtain metabolizing material: the main reason was the release of hydrolytic enzymes from the 'lysosomes' which decompose important co-factors such as ATP.

Warburg had used the slice technique solely for the degradative reactions which supply energy. My idea was to use it to study other metabolic processes, including synthetic ones, and as a first subject I chose the formation of urea in the liver. I was extraordinarily lucky in this choice because it led within twelve months to a major discovery—that of the ornithine cycle [32, 33, 34, 35], the first 'metabolic cycle' to be identified. A great many such cycles, perhaps more than a hundred, have been discovered since.

I was assisted in this work by Kurt Henseleit, one of the students who wanted to do research for the M.D. thesis. Although he had no special training in chemical work, or even in laboratory work generally, he learned the necessary techniques very quickly and proved to be a neat, hard-working, and reliable co-worker. After I had left Freiburg, he worked with Thannhauser for a short time. But there were no academic prospects for him in the Third Reich: his association with Thannhauser and myself made him *persona non grata* with the Nazis. He became a specialist in internal diseases and finally settled in Friedrichshafen on Lake Constance. I lost touch with him temporarily, but established regular contact again after the war.

I describe the development and significance of my work on urea synthesis in the next chapter.

Eventually, three decades later, my hope that this kind of work would become relevant to practical medicine was fulfilled. We now know of five diseases caused by a failure of the enzymes of urea synthesis. They are all inherited abnormalities of the various enzymes required for the synthesis of urea. The nature of the diseases is now understood, though any cure is still a long way off because the missing enzymes cannot be replaced within the cell. Dietary control can be helpful in cases where the abnormality is caused by a faulty enzyme (with weak activity) rather than by absence of the enzyme, i.e. mutation rather than deletion of a gene. If the enzyme is absent, the newborn baby cannot survive. If the enzyme is weak, the diet can be regulated (kept low in protein and taken in small quantities throughout the day) so that the enzyme is never overloaded and can cope with its function of detoxicating the ammonia arising from protein breakdown.

The discovery of the ornithine cycle was immediately recognized by the scientific community as a major achievement, which did much to improve my self-confidence. It led to invitations to lecture in Meyerhof's laboratory at Heidelberg, at the University of Frankfurt, and at Harnack House at Berlin-Dahlem, the social centre of the Kaiser Wilhelm Institutes. This invitation came from Max Planck, then President of the Kaiser Wilhelm Gesellschaft. Behind the invitation was Otto Warburg. It was he who arranged for the letter of invitation to be signed by Planck because he knew that this would please me, and because my paper had resolved his doubts about my competence. Carl Neuberg, a leading biochemist, put my name forward as a candidate for a chair of biochemistry, but this nomination was made only two weeks before the Nazis came to power. J. Parnas of Lwow, another leader in biochemistry at the time, asked me if I would have one of his collaborators, Dr Pawel Ostern, as a Rockefeller Fellow in my laboratory. But more than all the marks of recognition, what subsequently proved to be of crucial importance to me was the impression my work made upon Sir Frederick Gowland Hopkins, Professor of Biochemistry at Cambridge and President of the Royal Society of London. He quoted it as an important example of modern biochemical work and used it as the main scientific topic of his Presidential Address to the Royal Society on 30 November 1932 (2).

These responses to my research efforts were very encouraging. I clearly recall thinking, 'If this is what the world of science considers good work, then let me have the tools and the time and I will produce more of it.' Now I can say that I did so, but on looking back I also feel that at the time I did not give due credit to the large slice of luck which contributed to the discovery.

Luck was also on my side in that the discovery came when it did, one year before Hitler came to power. It established my reputation as an original investigator just at the time that I became forced to look for a post outside Germany.

After I had established the cyclic process of urea formation in slices, I made some efforts to obtain the reactions of the cycle in broken-up tissue, with the intention of studying the properties of the enzymes involved. These attempts were completely negative and I therefore turned to investigating other aspects of nitrogen metabolism. Some fifteen years later, Philip Cohen in Madison-Wisconsin (who spent a year in my laboratory at Sheffield in 1938) was successful where I had failed (3)—after much new knowledge on the handling of broken-up tissue had come to light and opened up new possibilities.

I decided to apply the tissue slice technique to the study of the degradation of amino acids [36, 37, 38]. This work led to the demonstration that the kidney is a major site of amino acid metabolism. It also led to the discovery of an enzyme (D-amino-acid oxidase) in liver and kidney which oxidizes the non-natural isomerides of the amino acids belonging to the D-series [48]. The oxidation occurs at the α-carbon atom and converts α-amino acids to the corresponding oxo-acids with the liberation of ammonia.

$$R.CH(NH_2)COOH + \tfrac{1}{2}O_2 \rightarrow R.CO.COOH + NH_3 + H_2O$$
(amino acid) (oxo-acid)

The widespread occurrence of this enzyme in the liver and kidney of vertebrates has been a puzzle ever since, despite the hundreds of papers which have been devoted to its study. Why should there be an enzyme which is highly active and yet is able to attack only substances which do not occur in significant amounts in the body or in food? To this day there is still no satisfactory answer to this question. Since characteristics which fulfil no useful function usually disappear in the course of evolution, one expects that D-amino acid oxidase does have a function, but so far no one has been able to discover what that function might be.

It was in Freiburg in 1932 that I began the first experiments which eventually led to the discovery of the citric acid cycle, which I describe in Chapter 9.

In December 1932 I was formally admitted as a teacher in Internal Medicine (*Privatdozent*), which entitled me to the privilege of lecturing but carried no salary. This required the approval of the Ministry of Education of the State of Baden. The contents of the confidential statement to the Ministry, dated 14 December 1932, from the Dean of the Medical Faculty, Professor E. Rehn, were made known to me twenty-four years after the event (see p. 17). It had been drafted by my chief, Professor Thannhauser. Some extracts (translated) are given below.

As an assistant physician in the Department of Medicine, Dr Krebs has shown not only outstanding scientific ability, but also unusual human qualities. He is an excellent doctor who has a profound understanding for the psychology of his patients,[6] and is willing to help them as a friend and guide. He is loyal and reliable and has made many friends.

At scientific congresses and at occasional lectures in the University, he has proved a skilful lecturer who can present his subject in an interesting and attractive way.

His recent scientific work, especially the paper on the synthesis of urea in the animal body, has established his international reputation. This paper is of fundamental importance and will be regarded in the future as one of the classics of medical research. Dr Krebs will therefore be a special asset to the teaching body of the Faculty.

Throughout my stay at Freiburg, the political situation in Germany was very confused and unstable because of the economic depression. Millions of people were unemployed. The political extremists, the Nazis and the Communists, were trying hard to exploit these difficulties but in Freiburg, at least, they seemed to get relatively little support. They were certainly not much in evidence until 30 January 1933, when Hitler became Reichskanzler. Then the situation changed dramatically.

5
THE DISCOVERY OF THE ORNITHINE CYCLE OF UREA SYNTHESIS
(1932)

This chapter is an account of the discovery of the ornithine cycle of urea synthesis [34, 296, 326], addressed to readers who have some grounding in chemistry. Other readers may bypass this chapter without loss of continuity. All that is needed for a proper understanding is an appreciation of the following seven formulae:

CO_2	NH_3	H_2O	$\begin{array}{c}NH_2\\|\\C=O\\|\\NH_2\end{array}$
carbon dioxide	ammonia	water	urea

ornithine	citrulline		arginine
CH_2NH_2	$\begin{array}{c}NH_2\\|\\C=O\\|\\CH_2NH\end{array}$		$\begin{array}{c}NH_2\\|\\C=NH\\|\\CHNH\end{array}$
CH_2	CH_2		CH_2
CH_2	CH_2		CH_2
$CHNH_2$	$CHNH_2$		$CHNH_2$
$COOH$	$COOH$		$COOH$

These last three substances are closely related; they differ only in respect to the upper section of the formulae.

I mentioned in Chapter 3 that when working in Warburg's laboratory I had been impressed by his 'tissue slice technique' (1), which had enabled him to make major discoveries in the field of cell metabolism and cancer. Slices of human or animal tissue 0.2 to 0.4 mm thick are thin enough to enable them to obtain adequate supplies of oxygen and nutrients by diffusion. (In living tissue these are supplied by circulating blood.) The slices contain 10 to 20 layers of cells, all of which (except those on the surface) are virtually intact. Using this technique, Warburg was the first to measure the normal rates of oxygen consumption and lactic acid production in many animal tissues.

I wanted to apply the technique to a synthesizing process and as I have said, I decided to study the synthesis of urea. Though a complex process, it was relatively simple in comparison with many other syntheses and its study is facilitated by the large quantities of urea produced by the liver. For example, a human adult produces 30 grams of urea in 24 hours on an average diet and three times as much on a high protein diet. This implies that it should be possible to measure the rate of synthesis accurately with very small amounts—say, 30 milligrams fresh weight—of liver tissue.

Before I could test the capacity of the slices of liver to synthesize urea I needed an analytical method that was quick, sensitive, and specific, and so I modified a method which had been developed earlier by Marshall (2) and van Slyke. The principle of this method is simple: in a slightly acid solution (pH 5.0), the enzyme urease is added to the solution to be examined. Urease converts urea into CO_2 and NH_3:

$$\underset{NH_2}{\overset{NH_2}{C}} = O + H_2O \rightarrow CO_2 + 2\,NH_3$$

The acid neutralizes the ammonia, and the amount of carbon dioxide appearing as gas is then measured manometrically (i.e. by the increase in gas pressure within a closed vessel). This allows the amount of urea in the sample to be calculated.

The method is rapid: a dozen determinations can be made in less than an hour, and very small quantities of urea (about 0.05 mg) can be determined to an accuracy of a few per cent.

Incidentally, the urease method introduced by Marshall in 1913 was one of the earliest analytical procedures to employ enzymes as analytical tools. Now there are several hundred enzymic methods of analysis.

My preliminary tests were promising. Urea formation occurred in slices and the rate increased on the addition of ammonium salts. But I was by no means sure whether the rates observed corresponded to those *in vivo*. When studying a biological phenomenon, it is always important to examine the whole process and not merely a fragment in a damaged tissue.

In order to provide optimum conditions for survival of the slices, I tried to devise a medium which simulated blood plasma as closely as possible in respect to its organic constituents. This meant modifying saline solutions of the type which Ringer (3), Locke (4), Tyrode (5), and Warburg (1) had introduced from the 1880s onward. Those of Ringer, Locke, and Tyrode were grossly deficient in two major plasma constituents, bicarbonate and carbon dioxide. The solutions had been arrived at mainly by trial and error, chiefly in searching for a medium which would maintain the activity of a frog's heart. Warburg included bicarbonate and carbon dioxide, but his medium lacked magnesium, phosphate, and sulphate, and the concentration of other ions still deviated from the physiological range. The idea behind the development of the new saline solution was my conviction that the concentrations of the various ions that are present in blood plasma (which are almost identical in all mammalian species and are very similar in other vertebrates) were not accidental; I believed they had evolved so as to be optimally attuned to the functions of the various organs. At that time it was already well established that the function of many organs depends upon the chemical composition of their environment. The new medium that I developed included magnesium, phosphate, and sulphate in physiological concentrations and has subsequently proved superior to all earlier saline plasma substitutes, not only for biochemical, but also for physiological and pharmacological work. It is now in general use.

This medium still differs from plasma in respect to colloid-osmotic pressure and the organic constituents of plasma such as glucose, fatty acids, amino acids, and hormones. These deficiencies can, however, be partly compensated by suitable additions.

In some situations an alternative to the saline medium is, of course, blood plasma, but this has disadvantages: its composition is not clearly defined and it contains a large number of metabolites such as glucose and urea, which make it difficult to measure any increase in these substances.

My tests with tissue slices suspended in the new medium showed that liver slices synthesize urea at rates closely comparable to those known to occur in the body. The experiments also showed that the rate of synthesis was greatly increased by the addition of ammonium salts or certain amino acids. After this encouraging beginning, I decided to measure systematically the rate of urea synthesis in the presence of a variety of chemical precursors, in the hope that the results might throw some light on the chemical mechanism of urea synthesis. I had no preconceived ideas nor any hypothesis about this mechanism, but I had in mind Warburg's work on cell respiration, during which he had studied the rate of cell respiration in the presence of cyanide, carbon monoxide, narcotics, and other substances, and had succeeded in drawing far-reaching conclusions about the properties of the enzymes of respiration.

Questions which suggested themselves to me for investigation included the following:

(1) Is ammonia an essential intermediate in the conversion of amino nitrogen to urea nitrogen? If it is, ammonia must yield urea at least as rapidly as do amino acids. If it is not, the rate of urea formation from amino acids might be more rapid than the rate from ammonia. (It became known more than 20 years later that half of the nitrogen required to form urea can be derived directly from amino acids, i.e. without passing through the stage of ammonia.)

(2) How do the rates of urea formation from various amino acids compare?

(3) Do substances suspected[7] to be intermediates, such as ammonium cyanate, behave like intermediates in that they can be converted into urea? The atoms of ammonium cyanate were known to be rearranged into urea in test-tube experiments:

DISCOVERY OF THE ORNITHINE CYCLE (1932)

$$N{\equiv}C{-}O{-}NH_4 \xrightarrow{\text{rearrangement}} NH_2.CO.NH_2$$
ammonium cyanate   urea

(4) Do pyrimidines, which are constituents of nucleic acids, yield urea directly or via ammonia? In pyrimidines the skeleton of urea bonds (N—C—N) is preformed.

In trying to answer these questions I measured the rate of urea synthesis under many different conditions, which included the presence of mixtures of ammonium ions and amino acids. It was in the course of these experiments that we discovered that urea was synthesized at exceptionally high rates when both ornithine and ammonium ions were present. The significance of this result was not immediately obvious, and it took a full month to reach the correct interpretation. At first I was sceptical about the validity of the observations. Was the ornithine perhaps contaminated with arginine? The answer was 'No'. Then it occurred to me that the effect of the ornithine might be related to the presence of arginase in the liver. Arginase is an enzyme which converts arginine into ornithine and urea; it has been known since the work of Kossel and Dakin (6) in 1904.

$$\begin{array}{c}
\diagup NH_2 \\
C{=}NH \\
\diagdown CH_2NH \\
| \\
CH_2 \\
| \\
CH_2 \\
| \\
CHNH_2 \\
| \\
COOH
\end{array}
\quad \xrightarrow{+\,H_2O} \quad
\begin{array}{c}
\diagup NH_2 \\
C{=}O \\
\diagdown NH_2 \\
\\
+ \\
\\
CH_2NH_2 \\
| \\
CH_2 \\
| \\
CH_2 \\
| \\
CHNH_2 \\
| \\
COOH
\end{array}$$

This represents a simple process by which urea is rapidly formed at the very same site where all surplus nitrogen is converted into urea, namely in the livers of all ureotelic animals, i.e. all of which excrete their surplus nitrogen in the form of urea; these include mammals, amphibians, and the reptiles belonging to the order Chelonia (turtles, tortoises).

Although the coincidence of the exceptionally high activity of arginase and the occurrence of urea synthesis had been known for decades, it had not occurred to anybody that there might be some connection between them, in the sense that the arginase reaction might be a component of the general urea-synthesizing mechanism. To me, after I had found the promotion of the synthesis of urea by ornithine, a connection seemed highly probable, though it was several weeks before I was able to formulate one. The solution to the problem developed only gradually as I studied the ornithine effect in detail.

In the first experiments which revealed the ornithine effect, the concentration of ornithine had been high because I had set out to explore whether ornithine acted as a nitrogen donor. When I tested lower ornithine concentrations, the stimulating effect remained. The final result of this aspect of the work was the discovery that one molecule of ornithine could bring about the formation of more than twenty additional molecules of urea, provided that ammonia was present. Moreover, the amino acid nitrogen content of the medium did not decrease during the synthesis of urea, while all the urea nitrogen could be accounted for by the disappearance of ammonia. This established the fact that ornithine acts like a catalyst: it is not used up and there is no simple arithmetical relationship between the amount of ornithine present and its effect upon the rate of urea production.

When considering the mechanism of this catalytic action I was guided by the concept that a catalyst must take part in the reaction and form an intermediate compound. The reactions of the intermediate must in this instance eventually regenerate ornithine and form urea. Once these postulates had been formulated it became obvious that arginine fulfilled the requirements of the expected intermediate. This meant that a formation of arginine from ornithine had to be postulated by the addition of one molecule of carbon dioxide and two molecules

DISCOVERY OF THE ORNITHINE CYCLE (1932)

of ammonia, and the elimination of water from the ornithine molecule.

It also became obvious at once that the synthesis of arginine from ornithine must involve more than one step, since four molecules—one ornithine, one carbon dioxide, two ammonia—had to interact. So I began to search for possible intermediates between ornithine and arginine. 'Paper chemistry' suggested that citrulline might play a role as an intermediate. The structural relations between ornithine, citrulline, and arginine are obvious from the formulae:

$$
\begin{array}{lll}
& \quad\quad NH_2 & \quad\quad NH_2 \\
& \quad\quad\diagdown & \quad\quad\diagdown \\
& \quad\quad C{=}O & \quad\quad C{=}NH_2 \\
CH_2NH_2 & CH_2NH & CHNH \\
| & | & | \\
CH_2 & CH_2 & CH_2 \\
| & | & | \\
CH_2 & CH_2 & CH_2 \\
| & | & | \\
CHNH_2 & CHNH_2 & CHNH_2 \\
| & | & | \\
COOH & COOH & COOH \\
\\
\text{ornithine} & \text{citrulline} & \text{arginine}
\end{array}
$$

By that time the occurrence of citrulline in biological material had just been established by two biochemists, working independently and on quite different topics. Wada (7) had isolated it from water melons (*Citrullus*) and Ackermann (8) had found it as a product of the bacterial degradation of arginine. I begged both of them for samples and obtained a few milligrammes from each, sufficient to do the decisive tests. The results entirely fulfilled our expectations: they demonstrated the rapid formation of urea in the presence of citrulline and ammonium salts in accordance with this scheme:

citrulline + ammonia → arginine + water $\xrightarrow{\text{arginase}}$ ornithine + urea

On the basis of these findings it became possible to formulate a cyclic process of urea formation from carbon dioxide and ammonia, with citrulline and arginine as intermediate stages, as shown in Fig. 1.

No cycle of this kind (called a 'metabolic cycle') had been known before—a cycle in which intermediate compounds of low molecular weight are formed cyclically. Entirely different kinds of cycle were of course familiar to biologists and chemists, such as the biological cycles like the menstrual cycle and the diurnal cycle. There is the carbon cycle, which entails photosynthesis in plants and combustion in animals. There are the life cycles of metamorphosing insects. There is the cell generation cycle. Closest to metabolic cycles are the catalytic cycles of chemistry, for example the actions of heavy metals in catalysing the 'knall-gas' reaction, the combination of hydrogen and oxygen to form water. This is catalysed by platinum or palladium and the catalysis involves the intermediate formation of the hydride of the catalyst.

Meyerhof (9) and the Coris (10) spoke of a 'lactic acid cycle' in muscle and liver. Lactate is produced from glycogen in muscle under anaerobic conditions during exercise and re-synthesized in the liver to glycogen. This is not strictly analogous to the ornithine cycle, but subsequently a large number of exactly analogous metabolic cycles has been discovered. Thus the ornithine cycle revealed a new pattern in the organization of metabolic processes.

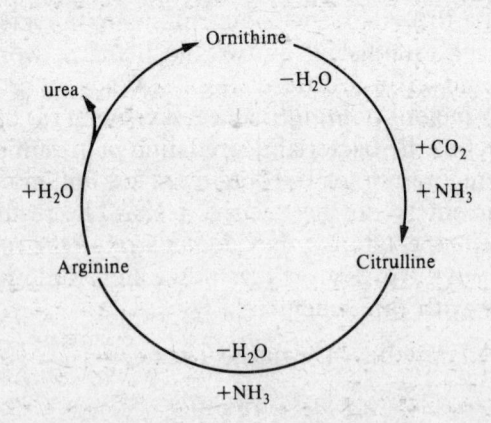

Fig. 1. Diagram of the ornithine cycle of urea synthesis.

DISCOVERY OF THE ORNITHINE CYCLE (1932)

The work I carried out in 1931 and 1932 established the outlines of the pathway by which urea is synthesized in the liver, but the time did not then seem ripe for attempting to unravel the enzymic mechanisms involved. A first step in that direction was to carry out experiments aimed at the synthesis of urea in cell-free material. Such experiments were negative in my hands, but where I failed, Cohen and Hayano (11) succeeded fourteen years later. By that time more had been learned about ways of preserving complex metabolic processes in cell-free homogenates and in tissue fractions. The development of techniques for preparing metabolically active cell-free homogenates was a lengthy endeavour in which many biochemists took part. One of the chief requirements of homogenates turned out to be the addition of soluble coenzymes and cofactors such as pyridine nucleotides, adenosine triphosphate (ATP), and magnesium. The concentration of these is diluted when cells are disrupted in an aqueous medium, and addition of these cofactors to the medium restores more or less normal concentrations. Work carried out by later investigators between 1946 and 1957 added several other intermediates, which the interested reader may find in textbooks of biochemistry.

The history of the development of the ornithine cycle illustrates the general experience that at any one time the solution of a problem can be advanced to only a limited extent. Soon seemingly impenetrable walls obstruct progress. After a time, however, advances in collateral fields overcome the barriers—sometimes by circumventing them, sometimes by demolishing them piece by piece.

While the majority of scientists commented favourably on the discovery, there were also adverse criticisms. In 1934, E. S. London (12), a Russian physiologist in Leningrad, argued that ornithine had no effect in his experiments on the isolated perfused dog liver. In 1942, Trowell (13) of the Cambridge Physiological Laboratory reported that the perfused rat liver did not respond to ornithine. At the same time Bach and Williamson (14) of the Cambridge Biochemical Laboratory claimed to have shown that dog liver can form urea from ammonia even when arginase is completely inhibited by high concentrations of ornithine, and they concluded that liver can synthesize urea without the participation of arginase. As late as 1956 Bronk and

Fisher (15) reported (from my own Department at Oxford) that under certain conditions citrulline is less effective than ornithine in promoting urea formation from ammonia and they suggested that 'ornithine and citrulline catalyse separate processes producing urea'. None of these criticisms has stood the test of time. They rested partly on inadequate techniques of investigation, partly on errors of interpretation.

By 1933 it had already been established that the ornithine cycle occurs generally in 'ureotelic' species. An interesting development was the demonstration by Srb and Horowitz (16) in 1944 that the mould *Neurospora* can synthesize arginine via ornithine through the reactions of the ornithine cycle, with citrulline as an intermediate stage. Later, many investigators showed that this is true for many microorganisms and for plants. The detoxification of ammonia in vertebrates thus utilizes reactions evolved earlier, for biosynthetic purposes, by very low levels of life.

The account of the discovery of the ornithine cycle illustrates the importance of new techniques to progress, especially those which make it possible to conduct a large number of experiments within a short time, and to study a process under many different conditions with a view to establishing factors which affect the rate of the process. It also illustrates the importance of following up an unexpected and puzzling observation arising in the course of experiments. Luck, it is true, is necessary, but the greater the number of experiments carried out, the greater is the probability of being lucky. Finally, the story shows that adverse criticisms are liable to be raised on the grounds that either the observations are not confirmed by other workers or that other observations do not fit in with the interpretation of the findings. Almost every major development in science meets with criticisms of this kind.

6
HITLER'S SEIZURE OF POWER
(1933)

Within a few days of Hitler's assumption of power Nazi uniforms appeared everywhere. Colleagues in the hospital who had admitted at most only mild sympathies with Hitler were suddenly to be seen in the uniforms of Nazi organizations. One wore three stars, indicating senior rank; many had been secret members for years. The Nazi press, political demonstrations, and broadcasts viciously aroused political feelings and intimidated opposition. What happened thereafter has become a matter of history.

For some eight weeks, until the end of March, work in the hospital and the laboratory went along almost normally, although the atmosphere was tense. A crucial day in the anti-Semitic campaign was 1 April 1933, when throughout Germany there was a boycott of Jewish shops.

On 12 April, I received a curt, formal, impersonal letter from the Dean of the Faculty, the Professor of Surgery, E. Rehn, officially informing me that at the request of the Minister of Education, I was to consider myself on immediate leave of absence:

Notification of immediate removal from office
By order of the Office of the Academic Rector I hereby inform you, with reference to the Ministerial Order A No. 7642, that you have been placed on leave until further notice.

Less than four months earlier this same Rehn had signed the letter which, in glowing terms, had recommended my appointment as a teacher in the Medical Faculty. The difference between the two documents demonstrates the breakdown of social tenets under ruthless political pressure—threats of being deprived of one's livelihood, of being discredited by public defamation, of being thrown into jail or a concentration camp, or of being murdered.

The Dean was implementing the following Government order:

The Minister of Education to the Administrative Director of the Freiburg University Hospital:
Maintenance of Security and Order
The Minister of the Interior has decided that all members of the Jewish race (irrespective of their religion) who are employed in the service of the State or in teaching establishments will be placed on leave of absence until further notice.

All academic teachers and assistants of this category are to be notified that they are placed on leave forthwith.

Should such leave cause immediate danger to patients, the leave may be delayed. In this case staff are obliged to continue their medical duties. The reasons for any delay are to be reported to me immediately.

This was followed on 18 April by official confirmation of my dismissal on 1 July 1933. The signature was that of the Administrative Director of the Hospital, a man I knew quite well. His son and I had recently collaborated on a paper [42] which was published on 22 April. Despite the father's earlier politeness towards me, the notification carried none of the customary opening and closing courtesies. It was ice-cold.

I am instructed by the Minister of Education to inform you that you have been relieved of your post in connection with the Law for the Reconstruction of the Professional Civil Service. Your contract will terminate on 1 July 1933.

You are requested to acknowledge receipt of this notification by your signature on the enclosed form and to return same immediately.
Eitel

The 'legal' instrument behind this letter, it will be noted, was called 'the Law for the Reconstruction of the Professional Civil Service'. Its hidden purpose was of course the removal of all non-Aryans and anti-Nazis. It resulted in the replacement of professionally highly competent people by others who were mediocre and sometimes incompetent. Such was the fabric of lies which pervaded the whole Nazi system and, alas, often misled the distant observer.

The last entry in my Freiburg laboratory notebook is dated 11 April 1933. My last experiments there, carried out between 1 and 11 April, revealed a new, exciting, and puzzling phenomenon. Slices of rabbit or guinea pig kidney removed large amounts of added ammonia in the presence of glutamate. I reported this finding in my last paper from the Freiburg laboratory, in the *Zeitschrift für Physiologische Chemie* [38]. This paper contained the sentence, 'I could not settle the question of which nitrogenous substances are formed from the added ammonia and glutamate because I was forced to break off my research'.

I found the answer later, in Cambridge: the product was glutamine [49]. This is synthesized from glutamate and ammonia if a source of energy is available.

This discovery was the first contribution to what was to become a large subject—that of the physiological functions of the amino acid, glutamine.[8] Up to that time, all that was known about glutamine was that it is a constituent of proteins; later many other roles were discovered. Glutamine is the main precursor of urinary ammonia and therefore plays a role in the regulation of the acid–base balance. The amide nitrogen of glutamine is the precursor of some of the nitrogen atoms of purines, pyrimidines, and amino sugars. Glutamine is a transport form of nitrogen set free in the degradation of amino acids in peripheral tissue to the viscera, especially the liver and the small intestine where the glutamine is further metabolized. In blood plasma and in many tissues, glutamine is the most common amino acid among the twenty amino acids occurring in animal tissues. The significance of its ready synthesis in kidney cortex is related to its multiple function. Several other amino acids are now also known to have multiple functions, for example arginine, whose role in urea synthesis I referred to in Chapter 5.

On 13 April I cycled to St Peter, a small resort in the Black Forest, ten miles from Freiburg. I stayed there for eight days to write up my recent work and await developments. The fierceness of Hitler's anti-Semitic policies which had hit me by my enforced 'leave' was much more severe than most of us had anticipated. It was perhaps wishful thinking to expect that Hitler's pronouncements before his assumption of power were, to a considerable extent, election propaganda. In practical politics, we believed, he would have to be much more moderate, partly because of internal popular opposition, partly because a radical elimination of the Jews would upset the economy and other aspects of German life, and partly because he would have to pay attention to world opinion.

News of the happenings in German universities had been arousing the concern of scientists abroad. A young German doctor, Walter Herkel, who had formerly worked at Freiburg and was then spending a year in Barcroft's laboratory at Cambridge, wrote to me from there on 8 April.

> I presume you will not take it amiss that, without your permission, I have today confidentially sounded out the field. I heard this morning of the misfortunes which you are sharing with many colleagues (whether the news is authentic I do not know) and I spoke immediately to Sir F. G. Hopkins, President of the Royal Society, who knows your papers and holds you in high esteem, as you may have already gathered from his address to the Royal Society last November. Whether something can be done is more than doubtful, mainly because of financial difficulties. He said he would 'be delighted to have you here', and will talk to the Vice-Chancellor within the next few days. But you must not hope for too much. If you could possibly make a start here on your own financial resources (living costs about 200–250 marks per month here) something better could be arranged later.
>
> I do not know what will become of me. I shall have to return to Germany next week. When they find out that my father, as a social democrat, has been dismissed they will presumably have no use for me.

In my answer I told Herkel that to the best of my knowledge no-one at Freiburg had yet been dismissed. 'Maybe', I wrote, 'things look blacker from a distance than they do to us here.' My letter continued,

> Nevertheless compulsory 'leave' is to be expected for some of us in the near future. It will be a preliminary measure, to be followed in most cases by notice of dismissal.
>
> Until this happens to me, I do not wish to take any definite steps. I had already thought of writing to Hopkins, knowing from his address to the Royal Society that he is interested in my work. I would therefore be most grateful if you could find out whether there is a possibility that Hopkins might accept me. I have enough money to keep me alive without earning for a few months, but not longer.

A few days later I received a letter from Albert Szent-Györgyi, written in Holland on 22 April, in response to my request some weeks earlier for a sample of ascorbic acid. In his reply he said he would send some, and then went on to write,

> I very much regret to hear that you are having personal difficulties in Germany. I was in Cambridge during the last few days, and people talked of trying to do something for you. I have of course encouraged them as much as possible, and I hope that my words will have contributed something towards the realisation of the plans.

In a handwritten postscript he added:

> If you really would like to go to Cambridge it would be best if you wrote to Hopkins and assured him that you would be content with a modest livelihood. There are no senior jobs and people might perhaps hesitate to offer you anything less.
> So if you would like to go to Cambridge, ask Hopkins for facilities to work. If you wish you may mention that I have encouraged you to write.

Naturally I took up Szent-Györgyi's suggestion and wrote to Hopkins by the next post.

Dear Professor Hopkins,
I have heard today from Professor Szent-Györgyi that your great kindness might make it possible for me to work at Cambridge. Professor Szent-Györgyi encourages me to ask you whether you can accept me in your laboratory.

As a Jew I am about to lose not only my present position, but any possibility of working at all in Germany. The Government has already forbidden me to work in the laboratory, although my salary is to continue for a few weeks.

I would consider myself most fortunate if I could continue my work in your laboratory. I will, of course, be content with a modest livelihood if it means that I can carry on with my research.

Would it perhaps suit you if I came to Cambridge in the near future to discuss matters with you? As I am prevented from working I have time to travel, and would consider it a great pleasure if you would allow me to present myself and to thank you personally for your sympathy.

Hopkins replied by return of post, on 29 April.

I admire your work so much that I am very anxious to help you.
If it should prove possible to obtain the necessary financial assistance, I shall be glad to find a place in this laboratory for you. There is

a movement in London to collect a fund for supplying maintenance in cases such as your own, and I have already sent your name, explaining that I should welcome you in this laboratory. It may take a little time, however, before I can let you know whether any money has been allotted.

Unfortunately Cambridge University itself is not able to find sufficient money to maintain the many applications that are being made.

I will let you know at the earliest opportunity what has happened with regard to the financial question.

Other offers of help arrived shortly afterwards. On 26 April Otto Warburg wrote:

I have just received an enquiry from Cambridge about details of your present situation, with a request for a prompt reply. It appears that people there want to do something for you if it becomes necessary.

Should you lose your position, I could probably accept you as a guest scientist and offer you a flat, a laboratory and about 250 marks a month. But England seems to me safer and more promising for you; we do not know what might happen here (although nothing has happened as yet in Dahlem).

On 30 April Franz Knoop wrote that he would be glad to help by offering me a place in his laboratory at Tübingen if this could be financed by non-government sources or by the Rockefeller Foundation, who were already supporting my research through a grant to Professor Thannhauser. I ought to mention here that Knoop and I had first got to know each other at the Thirteenth International Physiological Congress at Boston in 1929 and had travelled back to Germany together in a party of about twenty-four German scientists on the new German liner *Bremen*, when there were many opportunities for leisurely companionship. After this, we had corresponded over the publication of my papers in the *Zeitschrift für Physiologische Chemie*, the journal which he edited. We met again at the Fourteenth International Physiological Congress in 1932, this time in Rome, and found each other on the same northbound train when the Congress was over. We talked about our immediate plans and I told him that I was on my way to spend a week at a small seaside resort (Forte dei Marmi, near Pisa) where I would meet up with three or four colleagues, although my main object was to improve my

Italian. Knoop said that he would like to join us and I felt honoured because I respected him greatly and had found him always straightforward, unaffected, and friendly. Though he was twenty-five years my senior our relationship was on equal terms. So we had another pleasant few days together with plenty of time to talk.

I was also at this time following up a suggestion from Professor Thannhauser that I should get in touch with Professor Löffler at Zürich about the possibility of working in his department, the Medical 'Poliklinik' of Zürich University. I visited Löffler there and found him very sympathetic and interested because of his own earlier work on urea formation in the mammalian liver (1). He offered me a temporary place and told me that he could arrange, through private channels, a monthly salary of 350–400 francs plus some research money.

On 13 May Dr Robert A. Lambert of the Rockefeller Foundation wrote to me from Paris to enquire about my situation. I answered straight away and shortly afterwards he came to Freiburg. He gave me to understand that there was a distinct possibility, though by no means a certainty, that the Foundation would make me a maintenance grant sufficient to support me at Cambridge for one year. This would require further discussion with Professor Hopkins.

Some two weeks after his first letter Hopkins wrote again, in a very understanding vein:

I know you must be feeling impatient, but I feel I must ask you to wait a little before I can send any definite news of the situation.

The appeal for a large maintenance fund for workers from Germany has now been started but the issue is not yet known.

One of my staff had heard indirectly that they would welcome a biochemist at Oxford, but on communicating with my colleague, Professor [now Sir Rudolph] Peters, I found that this is, after all, uncertain. I had thought of suggesting to you that if we have not any room here you might like to come to Oxford where they have plenty of accommodation—but this matter also seems to involve a little delay.

I do not know whether you have had any other offers of hospitality; if not, I would ask you to wait a little for further information from myself.

After reading this, I wrote to Professor Peters at Oxford, referring to Hopkins's letter. I explained my position and enclosed reprints of my recent publications, but had received no reply by the time I left Freiburg on 19 June. When I visited Peters in Oxford at the end of June he explained that he had had nothing definite to tell me.

On 8 June I received another letter from Hopkins mentioning the Rockefeller Foundation's willingness to support me at Cambridge. He also referred to 'an approach from Oxford', about which I had heard nothing.

I am rather anxious to know whether you have personally received any approach from Oxford. I have heard from Dr. Lambert of the Rockefeller Foundation that they are willing to provide you with an income to work in this Department. I only hesitate to invite you straightaway because, being already committed to Dr. Friedmann and to Dr. Lemberg of Heidelberg (committed owing to previous circumstances), our accommodation has become rather limited and might be such as to disappoint you.

If Oxford has made any definite offer it might be better for you to go there.

Meanwhile the political atmosphere around me was deteriorating.[9] Posters addressed to German university students, bearing the following manifesto, appeared all over Freiburg. Much of the fanatic, mystic, weird, and hysterical quality of the German text is softened by translation, so a facsimile is reproduced as Fig. 2.

As part of The New Order:

AGAINST THE UN-GERMAN SPIRIT

1. Language and literature are rooted in the people. The German people bear the responsibility to ensure that their language and literature are a pure and uncontaminated expression of their national character.
2. Today a conflict is opening up a breach between the German literature and the German people. This is an outrage.
3. On you depends the purity of language and literature! Your Nation has entrusted its language to you for safe keeping.
4. Our most dangerous adversary is the Jew and he who is his vassal.
5. The Jew can think only as a Jew. If he writes German, he lies. The German who writes German but thinks un-German is a traitor. The student who speaks and writes un-German is, moreover, shallow and perfidious.

Im Rahmen einer Gesamtaktion:

Wider den undeutschen Geist

1. Sprache und Schrifttum wurzeln im Volke.
   Das deutsche Volk trägt die Verantwortung dafür, daß seine Sprache und sein Schrifttum reiner und unverfälschter Ausdruck seines Volkstums sind.
2. Es klafft heute ein Widerspruch zwischen Schrifttum und deutschem Volkstum. Dieser Zustand ist eine Schmach.
3. Reinheit von Sprache und Schrifttum liegt an dir!
   Dein Volk hat dir die Sprache zur treuen Bewahrung übergeben.
4. Unser gefährlichster Widersacher ist der Jude und der, der ihm hörig ist.
5. Der Jude kann nur jüdisch denken. Schreibt er deutsch, dann lügt er. Der Deutsche, der deutsch schreibt, aber un. deutsch denkt, ist ein Verräter. Der Student, der un. deutsch spricht und schreibt, ist außerdem gedankenlos und wird seiner Aufgabe untreu.
6. Wir wollen die Lüge ausmerzen,
   wir wollen den Verrat brandmarken,
   wir wollen für den Studenten nicht Stätten der Gedankenlosigkeit, sondern der Zucht und der politischen Erziehung.
7. Wir wollen den Juden als Fremdling achten, und wir wollen das Volkstum ernst nehmen.
8. Wir fordern deshalb von der Zensur:
   Jüdische Werke erscheinen in hebräischer Sprache.
   Erscheinen sie in Deutsch, sind sie als Übersetzung zu kennzeichnen.
   Schärfstes Einschreiten gegen den Mißbrauch der deutschen Schrift. Deutsche Schrift steht nur dem Deutschen zur Verfügung.
   Der undeutsche Geist wird aus öffentlichen Büchereien ausgemerzt.
9. Wir fordern vom deutschen Studenten Wille und Fähigkeit zur selbständigen Erkenntnis und Entscheidung.
10. Wir fordern vom deutschen Studenten den Willen und die Fähigkeit zur Reinhaltung der deutschen Sprache.
11. Wir fordern vom deutschen Studenten den Willen und die Fähigkeit zur Überwindung des jüdischen Intellektualismus und der damit verbundenen liberalen Verfallserscheinungen im deutschen Geistesleben.
12. Wir fordern die Auslese von Studenten und Professoren nach der Sicherheit des Denkens im deutschen Geiste.
13. Wir fordern die deutsche Hochschule als Hort des deutschen Volkstums und als Kampfstätte aus der Kraft des deutschen Geistes.'

Die Deutsche Studentenschaft.

Fig. 2. The manifesto 'Against the un-German spirit' published in the *Freiburger Studentenzeitung* 2 May 1933.

6. We will stamp out the lie, we will pillory treachery and treason, we demand for the students that the University shall be a place, not of shallowness but of discipline and political education.
7. We want to respect the Jew as an alien, and we want to take national character seriously.
   We therefore demand of the Censor that
   - Jewish works be published in the Hebrew language.
   - If published in German, it must be made clear that they are translations.
   - The most rigorous measures be taken against the misuse of the German language. Germans alone should be permitted to use the German language.
   - The un-German spirit must be stamped out in public libraries.
8. We demand from the German student the determination and ability for independent reasoning and decision.
9. We demand from the German student the determination and ability to preserve the purity of the German language.
10. We demand from the German student the determination and ability to overcome Jewish intellectualism and the threat it contains: the decay of the spirit of the German people through liberalism.
11. We demand that dedication to the safeguarding of the German

way of thinking must be the prime consideration in the selection of students and professors.

12. We demand a German University that is a guardian of the German National Character and a place of struggle springing from the strength of the German spirit.

The Students of Germany

The Rector of the University, von Möllendorff, Professor of Anatomy, ordered the removal of such posters from the University premises. He was immediately relieved of his office and was replaced by a man who did not object to the posters. This was the famous Martin Heidegger who had an international reputation as a distinguished philosopher and was widely regarded as a central figure in existentialist thought. He was an enthusiastic supporter of Hitler, hailing him on behalf of the University as a saviour, and pledging himself to unswerving loyalty.

Heidegger's lack of political acumen is illustrated by the following documents (2), only two of many similar writings. The German text of the manifesto contains newly-coined words and obscure, if not nonsensical, passages but at the same time it exudes a kind of high-minded mysticism. The original German text is therefore also reproduced.

German Students

The National Socialist revolution has brought about a catharsis of our German existence.

It is up to you in these times to be ever thrusting and prepared, ever tough and gaining strength.

Your urge to know seeks the Essential, the Simple, and the Great.

It is expected of you that you will seek out fellow thrusters and the most deeply dedicated, and expose yourself to their ideas.

Be hard and genuine in your demands.

Stay clear and certain about what you reject.

Do not let your new awareness be perverted into vain self-possession. Protect it as the indispensable foundation for the Führer-type man. You can no longer be mere listeners. You are committed to share not only knowledge but also action in the creation of the great future school of the German spirit. Everyone must first put his gifts and privileges to the test and then make them law. This will be achieved by the power of the fighting involvement in the struggle of the whole nation to find its true self.

Daily and hourly, let the faith of the will to follow grow stronger.

Without pause let your courage to make sacrifices grow, for the salvation of the essence and uplifting of the innermost force of our people in its land.
Do not let dogma and 'ideas' govern your existence.
The Führer, himself and alone, is the present and future German reality and law.
Strive to know more, ever deeper and deeper. From now on every thing demands decision, every action demands responsibility.
Heil Hitler!
Martin Heidegger, Rektor

Affirmation of Loyalty

The University, City and Students of Freiburg i.Br. have sent the following pledge of loyalty to the Reichs–Chancellor:
To the Führer, Berlin
To the Saviour of our nation in its crisis: from schism and dilemma to unity, resolution and honour; to the master and fighter for a new spirit of self-reliance in the community of the peoples, pledge unswerving loyalty the citizens, the students and the teachers of the university town in the farthest south-western German borderlands.

Freiburg Dr. Kerber, Oberbürgermeister
9 November 1933 von zur Mühlen, Führer der Studentschaft
Dr. Heidegger, Rektor.

DEUTSCHE STUDENTEN

Die nationalsozialistische Revolution bringt die völlige Umwälzung unseres deutschen Daseins.
An Euch ist es, in diesem Geschehen die immer Drängenden und Bereiten, die immer Zähen und Wachsenden zu bleiben.
Euer Wissenwollen sucht das Wesentliche, Einfache und Große zu erfahren.
Euch verlangt, dem Nächtsbedrängenden und Weitestverpflichtenden ausgesetzt zu werden.
Seid hart und echt in Euerem Fordern.
Bleibt klar und sicher in der Ablehnung.
Verkehrt das errungene Wissen nicht zum eitlen Selbstbesitz. Verwahrt es als den notwendigen Urbesitz des führerischen Menschen in den völkischen Berufen des Staates. Ihr könnt nicht mehr die nur «Hörenden» sein. Ihr seid verpflichtet zum Mitwissen und Mithandeln an der Schaffung der künftigen hohen Schule des deutschen Geistes. Jeder muß jede Begabung und Bevorzugung erst bewähren und ins Recht setzen. Das geschieht durch die Macht des kämpferischen Einsatzes im Ringen des ganzen Volkes um sich selbst.

Täglich und stündlich festige sich die Treue des Gefolgschaftswillens. Unaufhörlich wachse Euch der Mut zum Opfer für die Rettung des Wesens und für die Erhöhung der innersten Kraft unseres Volkes in seinem Staat.
Nicht Lehrsätze und «Ideen» seien die Regeln Eures Seins.
Der Führer selbst und allein ist die heutige und künftige deutsche Wirklichkeit und ihr Gesetz. Lernet immer tiefer zu wissen: Von nun an fordert jedwedes Ding Entscheidung und alles Tun Verantwortung.

Heil Hitler!
Martin Heidegger, Rektor

TREUEKUNDGEBUNG

Universität, Stadtverwaltung und Studentenschaft von Freiburg i. Br. haben folgende Treuekundgebung an den Reichskanzler abgesandt:
An den Führer, Berlin.
Dem Retter unseres Volkes aus seiner Not, Spaltung und Verlorenheit zur Einheit, Entschlossenheit und Ehre, dem Lehrmeister und Vorkämpfer eines neuen Geistes der selbstverantwortlichen Gemeinschaft der Völker versprechen unbedingte Gefolgstreue die Bürgerschaft, die Studentenschaft und die Dozentenschaft der Universitätsstadt in der äußersten südwestdeutschen Grenzmark.
Freiburg, 9. November 1933.

Dr. Kerber, Oberbürgermeister.
von zur Mühlen, Führer der Studentenschaft.
Dr Heidegger, Rektor.

On 15 July 1933, Heidegger put his signature to the document which terminated my appointment as a university teacher.

The Minister of Education and Justice rules as follows:
According to preliminary investigations, the Law for the Reconstruction of the Professional Civil Service of 7 April 1933, applies, according to Paragraph 3, to Dr. H. A. Krebs. He is asked to comment on this within three days.
You are hereby notified of this, and should present your comments, if any, not later than 10 a.m. on 20 July 1933 at my office.

There was appalling confusion in the minds of educated Germans. One example is the following letter I received from a young medical graduate whom I had accepted as a research worker in my laboratory before Hitler became Reichskanzler but later had to inform that political events had made it impossible for me to keep to the agreement.

I greatly regret the necessary cancellation of my acceptance in your laboratory. You know clearly enough from my intended visit how highly I esteem your work. So I trust that you will receive, in this same spirit, the warmest and widest hospitality from my people, as has always been our noble custom.

If the new laws, necessitated by certain malpractices, are affecting you in such a regrettable way, I ask you to bear solely in mind the will of the Führer, who has always shown the noblest striving to avoid unjust hardships, and to overlook your own experience as I willingly forget the pogroms which the Jews held it necessary to organise against German tourists in Tunis.[10] I hope that after the confusion of the present reorganisation there will be a happy settlement in your own case and that perhaps I may still have an opportunity to visit you.

Sehr geehrter Herr Dozent Dr. Krebs!
Die notwendige gewordene Absage meiner Aufnahme in Ihren Laboratorium bedauere ich ausserordentlich. Sie wissen aus meinem vorgehabten Besuch eindeutig genug, wie hoch ich Ihre Arbeiten einschätze. So hoffe ich auch, dass aus einer gleichen Einstellung heraus Ihnen die herzlichste und weiteste Gastfreundschaft in meinem Volke gewährt wird, wie sie stets edelste Sitte gewesen ist.

Wenn durch gewisse Misstände veranlasste gesetzliche Regelungen im Augenblicke auch Sie in so bedauerlicher Form treffen, so bitte ich Sie, lediglich an den Willen des Führers zu denken, der in edelster Gesinnung stets sein Bemühen gezeigt hat, ungerechte Härten ganz zu vermeiden, und Ihren eigenen Vorfall nachzusehen, wie auch ich gern die Pogrome vergesse, die die Juden in Tunis gegen die deutschen Reisenden veranstalten zu müssen glaubten. Ich hoffe, dass nach der Unklarheit der augenblicklichen notwendigen Umstellung in Ihren Falle die erfreulichste Regelung Platz greift und dass ich Sie gelegentlich doch aufsuchen kann.
Mit besten Gruss
Ihr sehr ergebener

The great majority of the ordinary people, including university staff, accepted the Nazi policies as something they could do nothing about. Many tried to justify their behaviour to themselves by emphasizing the allegedly good features in the changes and explaining away the excesses and outrages as but temporary unpleasantnesses which would disappear with time.

People with genuine moral courage were very rare. A relatively outspoken critic was Ludwig Aschoff, the most distinguished member of the Freiburg medical faculty. Because of his high standing, both in the university and abroad, the Nazis had to bear with his comments. He was strongly opposed to their anti-Semitic policies and had the courage to demonstrate this openly. On May Day 1933 all university staff had to take part in a great parade under the Nazi flags at eleven o'clock. Jewish and 'non-Aryan' members were excluded. Twenty years later, I learned that Aschoff had visited several of his Jewish colleagues on that day in order to express personally his disapproval of the government's anti-Semitic policy. Every member of his family made a similar gesture. On the night before the parade I myself received a telephone call from his daughter Eva, whom I had met at a few social occasions. She asked me if I would go for a walk with her at eleven o'clock the following morning. I understood at once what she had in mind, and accepted.

Some of the professors were quite intimidated. One was told by a member of his staff that he intended to refuse Nazi orders to join the storm troopers, the para-military organization. Joining meant regular attendance on Wednesday afternoons, and the young man told his chief that he proposed to work in the laboratory instead. The professor tried very hard, but unsuccessfully, to persuade him to comply with orders, and when the young man insisted, told him that the laboratory would be closed on Wednesday afternoons.

How easily academic institutions of high standing succumbed to Nazi pressure is forcefully brought home by the correspondence between the Prussian Academy of Sciences and Albert Einstein (3). Einstein was at that time one of the most famous and most distinguished German scientists—if not the most famous and most distinguished. He happened to be abroad when the Nazis came to power. On 1 April 1933 the Prussian Academy issued the following declaration:

> The Prussian Academy of Sciences heard with indignation from the newspapers of Albert Einstein's participation in atrocity-mongering in France and America. It immediately demanded an explanation. In the meantime Einstein has announced his withdrawal from the Academy, giving as his reason that he cannot continue to serve the Prussian state under its present government. Being a Swiss citizen, he

also, it seems, intends to resign the Prussian nationality which he acquired in 1913 simply by becoming a full member of the Academy.

The Prussian Academy of Sciences is particularly distressed by Einstein's activities as an agitator in foreign countries, as it and its members have always felt themselves bound by the closest ties to the Prussian state and, while abstaining strictly from all political partisanship, have always stressed and remained faithful to the national idea. It has, therefore, no reason to regret Einstein's withdrawal.

It should be noted that this was written after the general boycott of Jewish shops on 1 April. One surmises that the great majority of the members of the Academy could not have agreed or approved of this statement but felt unable to do anything about it. There followed an exchange of letters in which Einstein refuted the charge of atrocity-mongering and upheld his resignation on the grounds of his belief in the freedom and equality of the individual. The Academy, however, rebuked him for withdrawing his support and for doing 'nothing to counteract unjust suspicions and slanders', which, in the opinion of the Academy, it was his duty as one of its senior members to do.

We had confidently expected that one who had belonged to our Academy for so long would have ranged himself, irrespective of his own political sympathies, on the side of the defenders of our nation against the flood of lies which has been let loose upon it. In these days of mud-slinging, some of it vile, some of it ridiculous, a good word for the German people from you in particular might have produced a great effect, especially abroad. Instead of which your testimony has served as a handle to the enemies not merely of the present Government but of the German people. This has come as a bitter and grievous disappointment to us, which would no doubt have led inevitably to a parting of the ways even if we had not received your resignation.

A telling illustration of the horrible atmosphere created by the Nazi philosophy is the following story, which was sent to me by a friend in 1937. It concerned my former colleague and friend, Dr Arthur Jores and my former chief at Altona, Professor Leo Lichtwitz. It appeared in the daily *Niederdeutscher Beobachter*. The arrogant and evil style of the article and the malicious denunciation by a colleague, are characteristic of the Nazi mentality. Dr Jores was a man of great integrity and great ability. At the time of the incident he was a *Privatdozent* (teacher) in medicine at Rostock University. Professor Lichtwitz, his former chief and teacher, had been dismissed from his post as a 'non-Aryan' and thereby forced to leave Germany. Because of his outstanding reputation he was soon appointed to a post in New York. In 1936 Dr Jores sent him a reprint of one of his publications with a personal dedication—*'meinem verehrten Lehrer'*. This was intercepted by a Nazi colleague who saw the letter ready for posting in the departmental office, took it away and promptly denounced his colleague to the party.

As a result, Dr Jores was deprived of his position at Rostock; an academic career and many professions were closed to him thereafter. He found refuge as a 'pharmacologist' in a pharmaceutical firm, Promonta, at Hamburg, thanks to the understanding of its directors. The Nazis, however, continued to persecute Dr Jores as a politically unreliable person. He was imprisoned and kept for a long time in solitary confinement.

Soon after the war, in 1946, when the occupying powers were seeking men with a clean record to fill leading positions, he was reinstated in an academic appointment as Professor and Director of the Department of Internal Medicine at the University of Hamburg. In 1950 he became its Rector.

The writer of the article in *Niederdeutscher Beobachter* argued that 'loyalty' to a Jewish teacher was a crime in the Nazi Reich because, as a Jew, he must be an enemy of the German people. In fact, Lichtwitz was one of the leading German clinicians of his age, with an excellent record of achievement and integrity. He was an idealist, and a great patriot by all reasonable standards, though not by the distorted standard of the Nazis.

On Loyalty
by Friedrich Schmidt

A teacher in the University of Rostock, Dr Jores, tried to send one of his scientific papers with a personal dedication to a Jew in New York. This Jew played an important role in Germany before Adolf Hitler came to power, but after his accession, he promptly shook the dust of Germany from his feet—like many other former citizens with a bad conscience.

The Jores case interests us little in respect of the fate of this teacher in the National Socialist State, because we know that the authorities will deal appropriately with the matter. For us it is much more important as an instructive and telling example.

Dr Jores sees in the emigré Jew an honoured teacher to whom he owes a great deal. To explain his action, Dr Jores would say something like, 'Loyalty between man and man is a thing of value'. Of course we agree with such a statement—provided the loyalty is not extended to a man who has dissociated himself from the Nation and who has joined a group of people fundamentally opposed to this Nation, who attack us in base and malevolent ways.

Here loyalty confronts loyalty.

Only a man with a very strange attitude can see a conflict of loyalties in this situation—only a man who can visualize one individual in isolation from the rest and who holds his bond to another individual as being equal to, if not above, his bond of loyalty to his country. This is a kind of insanity in badly educated intellectuals, for which there are effective and thorough remedies.

By dedicating his paper to a Jew, Dr Jores (a German university teacher who was sworn to the education of German youth in National Socialist ideals) has confessed—according to the standards of any unprejudiced, clear-thinking German—that he is friendly towards an enemy of his country. He admits that he is against his country, against his duties, against the ideals of the German university and against German youth. Doubt is impossible. The confession is patently clear.

In the University of Rostock there have been objections and indignation at the fact that the case of Dr Jores has been made public by the Party and that his behaviour has been exposed to general censure. This reaction of University people to the question of unconditional bondage to the Nation demonstrates that the road to a true National Socialist University is still a very long one. These circles in the University belong to the 'other side' in the struggle of National Socialism to bring about a new interpretation of all scientific work, aimed at integrating each individual effort into the whole, at directing each individual effort systematically into the service of the Nation, in a great, all-embracing concept—the concept of the Nation as the task imposed upon each individual by God.

The 'other side' still wants to live in a world which has ceased to exist, and thereby deceives itself and us. But we will uphold the new and old and eternal truth of the holy law of nations in the face of this deception. We will choose even the ultimate sacrifice rather than allow the dead world, which was always a lie, a perversion of the truth, to gain power over us once more. These men on the 'other side' must step down if they do not wish to serve; they must make way for life and truth, just as the night makes way for the sun.

The documents I have quoted demonstrate the collapse of all accepted standards of decency in public life. The old standards

Ein Kapitel „Treue"

Von Friedrich Schmidt

Ein Dozent der Universität Rostock Dr. Jores versuchte ein von ihm verfaßtes wissenschaftliches Werk mit einer persönlichen Widmung an einen Juden nach Newyork zu senden, der bis zur Machtübernahme in Deutschland eine große Rolle gespielt hat und nach der Machtübernahme durch Adolf Hitler den deutschen Staub von den Füßen schüttelte, also Emigrant wurde — wie so viele andere ehemalige deutsche Staatsbürger mit schlechtem Gewissen.

Dieser Fall Jores interessiert uns weniger nach der Richtung hin, wie er sich im nationalsozialistischen Staat für diesen Dozenten auswirken wird, denn wir wissen, daß die zuständigen Stellen das Entsprechende veranlassen werden. Für uns ist dieser Fall vielmehr als hervorragendes Lehrbeispiel wichtig.

Dr. Jores sieht in dem ausgewanderten Juden einen verehrten Lehrer, dem er viel verdankt. Er sagt zur Erklärung seiner Handlungsweise etwa: „Die Treue von Mann zu Mann ist auch etwas wert!" Nun, dieser Satz ist selbstverständlich auch von uns anzuerkennen. Voraussetzung ist dabei allerdings, daß diese Treue nicht einem Mann bewiesen wird, der sich von der Nation des die Treue beweisenden Mannes losgelöst und zu einer Menschengruppe gestellt hat, welche diese Nation bis aufs äußerste, auch mit den niederträchtigsten Mitteln bekämpft.

Hier steht Treue gegen Treue.

Aber nur ein Mensch mit ganz besonderer Einstellung kann hier in einen inneren Konflikt geraten, eben nur ein Mann, der den Einzelnen losgelöst vom Ganzen sich denken kann und die Bindung zu einem andern Einzelnen der Bindung an das Volksganze überordnet oder mindestens gleichsetzt. Eine Seelenkrankheit falsch erzogener Intellektueller, gegen die es durchaus wirksame Mittel gibt.

Mit seiner Widmung an den Juden drüben bekennt sich — nach der Auffassung jedes unvoreingenommenen, klar denkenden deutschen Mannes — der deutsche Dozent Dr. Jores, der deutsche Jugend zu volksbewußten, nationalsozialistischen Männern zu erziehen gelobt hat, für einen Feind seines Landes, gegen sein Volk, gegen seine Aufgabe, gegen den Sinn der deutschen Hochschule, gegen die deutsche Jugend. Ein Zweifel ist unmöglich. Das Bekenntnis ist völlig eindeutig.

In Kreisen der Universität Rostock hat es Unwillen, ja sogar Empörung ausgelöst, daß der Fall Dr. Jores von Parteiseite her öffentlich behandelt und die Haltung dieses Dozenten der allgemeinen Mißbilligung preisgegeben wurde. Die Antwort dieser Universitätskreise auf die Frage an ihre bedingungslose Volksverbundenheit beweist uns, daß der Weg zur wirklich volksverbundenen, nationalsozialistischen Hochschule noch sehr weit ist. Diese Kreise bilden auch die **andere Seite** im Kampf des Nationalsozialismus für eine neue Sinngebung aller wissenschaftlichen Arbeit, die Einzelleistung auf das Ganze hinzulenken und einzubauen in das System des Dienens in einer großen, alles umspannenden Idee, der Idee des Volkes als der von Gott jedem Menschen gestellten Aufgabe.

Die „andere Seite" will immer noch in einer Welt leben, die nicht mehr vorhanden ist, und betrügt so sich selbst und uns. Wir aber stellen die neue und alte und ewige Wahrheit vom heiligen Volksgesetz des Menschen ihrem Betrug entgegen und wollen lieber das Letzte einsetzen, denn jene gestorbene Welt, die immer Lüge gewesen ist, jemals wieder eine Macht über uns werden zu lassen. Diese Menschen von der „anderen Seite" müssen abtreten, wenn sie nicht dienen wollen, und dem Leben, der Wahrheit weichen, wie die Nacht der Sonne weicht.

Fig. 3. 'On loyalty', an article denouncing Dr Jores, which appeared in the *Niederdeutscher Beobachter* in 1937.

were replaced by outrageous lies, lawlessness, personal vituperation, violence, torture, and murder. Under the banner of nationalism, a group of madmen, with fanatical and mythical slogans, gave their evil politics the appearance of a kind of holy idealism. It was not the first time in the history of mankind that such distortions of truth have happened. One hopes, with some trepidation, that it will have been the last. That so many Germans—and others, outside Germany—allowed themselves to be deceived by Nazi demagogy is an expression of extraordinary political immaturity. (The 'others' included at least one member of the British Government. In March 1938, The Marquess of Londonderry, who had been Secretary of State for Air until 1935, published his book, *Ourselves and Germany* (4). In it he quoted his letter to Ribbentrop of February 1936 referring to his 'charming interview with the Führer', which had lasted over two hours. He was completely taken in.) I presume that the majority of the German people never realized what they were bringing upon themselves when they voted for the Nazi Party, and that later, after Hitler had assumed absolute power, they were helpless in the face of the Nazi terror. My ultimate faith in the German people was, however, never shaken. I knew that many despised and hated the Nazis and that many paid with their lives for attempting to resist their brutal policies.

Although by the middle of June I had not received a final decision from Cambridge, I decided to leave for England as soon as possible. I knew from Hopkins's letter of 6 June that the Rockefeller Foundation was willing to support me for a while. I had had to choose between Cambridge and Zürich. Professor Löffler, whom I had kept informed about my correspondence with Hopkins, assured me very warmly that although he would greatly welcome collaboration with a colleague who was biochemically orientated, he also appreciated my interest in Cambridge. One consideration in favour of Zürich was the possibility of remaining closely associated with clinical medicine, in which I was deeply interested. But the warmth of Hopkins's letters, the great reputation of Cambridge and above all my feelings for England, stemming from my earlier contacts with the country and its people, drew me to Cambridge—a decision I have never regretted.

Fortunately Nazism had not yet penetrated every branch of the administration and I obtained permission to take with me to Cambridge most of the scientific instruments I had bought either with the Rockefeller grant or with funds provided by the Government *Notgemeinschaft*, a body comparable with the British Research Councils. I convinced the sympathetic officials that the work for which the apparatus had been purchased was not yet completed and that I would be finishing it at Cambridge. The apparatus and chemicals were packed into sixteen wooden boxes and several suitcases; they included two whole Warburg baths with twenty-four manometers. The administrators and the police authorities were happy to let the apparatus go, but several of my colleagues in the hospital whispered that I was filling sixteen boxes without any check being made on their contents. As a matter of fact, since I was not allowed to enter the laboratory, the boxes were packed by my loyal technicians. This equipment was of enormous help to me in Cambridge and also later in Sheffield.

Circumstances made it impossible for me to visit Hildesheim to say farewell to my family. I had last seen them in December 1932 when, after my lecture in Berlin-Dahlem, I made a detour in order to visit my father, his second wife, Maria Werth, a teacher whom he had married in 1931, and my ten-month-old half-sister, Gisela. It was to be my last visit.

My father was very sad, even distressed, that circumstances forced me to leave Germany. He expressed his feelings in words of special solemnity in a letter a few days before my departure.

Mein lieber Hans!
Bevor Du den deutschen Boden verlässt, möchte ich Dir noch herzliche Grüsse und Wünsche zusenden. Möge im Ausland Dir noch mehr Anerkennung und Glück zufallen, als es Dir bisher in der Heimat beschieden war! Ich hoffe, dass wir uns nicht seltener wiedersehen werden, als bisher, sei es in England, sei es in Deutschland. Ich fange sogar an, etwas englisch zu lernen (bei Maria), trotzdem meiner 66-jährige Zunge die fremde, seltsame Aussprache nicht leicht fällt. Recht herzlich bitte ich Dich, öfters zu schreiben als bisher und auch ausführlicher. Insbesondere teile doch auch bald mal mit, wie eigentlich die Stelle beschaffen ist, welche Du erlangen wirst, ob sie eine Dozentur darstellt oder eine Forschungsgelegenheit, Gehalt, Aussichten usw.

Von uns ist nicht Neues zu melden. Auch von den Berlinern Geschwistern nicht. Einer Notz in der Frankfurter Zeitung glaube ich entnehmen zu müssen, das die 'Reinigung' der AEG Beamtenschaft vielleicht noch nicht beendet ist.

Zum Schluss noch Eines: Wenn es Dir, was ich herzlich wünsche und erhoffe, im Ausland gut geht, vergiss nicht Deine in Deutschland lebende Geschwister! Vor allem nicht die kleine Gisela und auch nicht deren Mutter, denen Du vielleicht Schutz und Stütze zu werden berufen sein wirst.

Ich umarme Dich innigst
Vater

(Translation)
My dear Hans,

Before you leave German soil, I want to send you warmest greetings and wishes. Would that you will find abroad still more recognition and good fortune than has been accorded you in your homeland. I hope that we shall see each other not less frequently than hitherto—be it in England, be it in Germany. I have even started to learn some English (from Maria) although my 66-year-old tongue does not find it easy to cope with the unfamiliar and peculiar pronunciation. Very earnestly I beseech you to write more frequently than you have in the past, and also in more detail. Especially let me know soon exactly what the position is that you are going to. Is it a teaching post or a research appointment? What will the salary be and what are your prospects?

There is nothing new to report from us, nor from [Lise and Wolf in] Berlin. There was a comment in the *Frankfurter Zeitung* which seemed to indicate that the elimination of Jews from the AEG [General Electric Company where Wolf was employed] has not finished yet.

One final point: If you do well abroad, which is my sincere wish and hope, do not forget your brother and sister in Germany, and above all little Gisela and her mother. It may be that you will be called upon to be their shield and succour.

I embrace you most affectionately

Behind my father's sadness was his love for the Germany of his life and times. He regarded himself as a true patriot, in the best sense of the word, deeply attached to German civilization—its literature, music, fine arts, sciences, and scholarship. He admired the idealized virtues personified by Frederick the Great and Bismarck: self-discipline, a sense of duty, reliability, integrity, candour, diligence, tolerance, fairness, and moral and physical courage, and he cared very much that his children should grow to love the same high principles. Thus it was sad for him that I was leaving Germany.

On 19 June I said goodbye to Henseleit and my other immediate collaborators on my way out of the hospital. It was a sad and deeply moving moment. Then I made my way alone to Freiburg station, to catch the eleven o'clock train.

7
CAMBRIDGE (1933–1935)

In Strasbourg I changed to a through train to Dunkirk, where late at night I embarked for Folkestone. I vividly remember approaching the white cliffs of Kent at dawn, between four and five o'clock, wondering what would be in store for me in England. I was neither sad nor sentimental, rather I was full of curiosity and optimism, thanks to the spirit of Hopkins's letters and the promised support from the Rockefeller Foundation. Optimism—in the sense of hoping for the best although preparing for the worst—has always been my keynote; instinctively I tend to fasten on to favourable aspects and play down adverse ones, and to act accordingly—in everyday life, in planning research, in my assessment of people and in my dealings with them. And I have always felt sorry for those who cannot help but emphasize difficulties and play down positive opportunities.

I arrived in England almost penniless although I had a bank account in Germany of about 1500 marks, equivalent at that time to about £100. Of this, 1000 marks had been a gift from my father in December 1932 on the occasion of my admission as a university teacher. The strict German currency regulations did not allow emigrants or ordinary travellers to take more than 10 marks (less than £1) out of the country. In addition, another 10 marks could be transferred abroad once a month. I was allowed, however, to take personal effects with me, including all my books. At some risk I smuggled German banknotes worth about £20, putting them between pages of books, stuck together at the edge. The money I brought to Britain was sufficient to maintain me for several weeks, and I was comforted by the feeling that I possessed one great asset: I was well-trained in a profession which was internationally recognized.

The boat train reached Victoria at a quarter to eight in the morning on 20 June and I was met by my long-standing friend Hermann Blaschko, who had left Germany as a refugee seven weeks earlier. He was staying with an aunt in London, Mrs Helene Nauheim of Connaught Square, and offered me hospitality on her behalf, which I was glad to accept. Immediately after my arrival I wrote to Professor Peters at Oxford to say that I would be grateful if he could see me, referring again to my letter from Hopkins. I wrote to Hopkins also, informing him of my arrival in London, and asking for an interview.

I travelled to Oxford on 23 June. It was a sunny day and I was greatly impressed by the view of its spires from the window of the train. During my talk with Peters, he explained that the possibility of my working in his Department depended entirely on Balliol College, which might, or might not, have a lectureship for me. Peters himself had no position to offer, but if Balliol would give me a post, he would find a place for me in his laboratory. He entertained me to tea with his senior staff in the room which, twenty-one years later, was to become my office when I succeeded him. At this meeting I was introduced to Cyril Carter and Percy O'Brien, both of whom were later to become my colleagues. After tea Peters took me to Balliol College to meet the Master, A. D. Lindsay. This interview ended inconclusively and left me rather puzzled. I was told that I would hear from them later but Lindsay had seemed rather remote and I could not help feeling that no-one was particularly keen to have me. (My impressions proved incorrect as I was to learn, too late, a few days afterwards, in a letter from Peters.)

The next day I received a reply from Hopkins saying he could not see me during the week because he was committed to spending a lot of time in London on Royal Society business. I think he appreciated my anxiety, and so invited me to call at his home on Sunday afternoon. So I went to Cambridge on Saturday 24 June. My first aim was to find a room in a modest hotel and I booked in at the 'Glengarry' in Regent Street. I then set off for the Biochemical Laboratory. On my way I met K. A. C. Elliott, whom I had got to know the previous year at a symposium in Heidelberg. He introduced me to several people, among them David Keilin, whom I had also met at Heidelberg. Keilin's first question after greeting me was to ask about my lodgings and when I told him that I had a hotel room he expressed regret that I was not staying at his house. I was charmed and touched by this friendliness, but felt I could not very well cancel my booking, so we arranged that I should move to his house in Hills Road on the following day.

I went to Hopkins's house in Grange Road early on the next afternoon. On Sunday afternoons his house was always open to callers and several people came to tea on that day, among them Barbara Holmes, his daughter, with whom I had corresponded about the nitrogen metabolism of the kidney. I was warmly welcomed by everyone there, especially by Barbara who expressed keen interest in my coming to Cambridge because of our joint interests. Hopkins reaffirmed that he would be glad to have me but said he hesitated to press the matter because of the 'low' salary—the £300 per annum offered by the Rockefeller Foundation—and the temporary nature of the post. He said that a person of my standing should not be offered 'so little' but in the circumstances it was all he could offer. In fact, it compared quite well with my pay and allowances in Freiburg. In view of the uncertainty of the prospects at Oxford I accepted without any further thought. The working conditions seemed suitable, especially since I had brought basic equipment with me. The warmth of my welcome and his obvious keenness to have me was decisive. I returned to London on Monday, 26 June to prepare to move, and informed Peters of my decision. A day or two later I returned to Cambridge, where I had been invited to stay with the Keilins while I looked for lodgings. These I found at 46 Hills Road. I started experimental work in the laboratory on 7 July—two and a half weeks after leaving Freiburg and twelve weeks after being expelled from my laboratory—a very brief interruption considering the circumstances.

On 28 June I received the following letter from Peters:

Your letter came this morning, and also one from the Master of Balliol, which I enclose. Although you have decided to go to Cambridge, I thought that you had better see it so that you would understand the efforts which have been made to accommodate you here. The offer of Balliol College would of course have made it possible for us to have you in the laboratory. If you cared to interview me again, I shall be in London tomorrow (Wednesday) examining, at King's College (Strand) and could see you there.

I am glad that your affairs are settling themselves so satisfactorily.

The enclosed letter from the Master of Balliol read:

I have been talking with my scientists, and I think it is clear that if Krebs got a Rockefeller grant and should work with you, (a) we could get his laboratory expenses paid from the Jewish fund, and (b) the College would be glad if he would give us a little teaching—about as much as would at the ordinary rates come to £100 a year. This, you will understand, would only be for one year, but we should be very glad to do that.

I settled in Cambridge very quickly. Thanks to the equipment I had brought with me my facilities were very good; right from the start I had all the special apparatus I needed, such as Warburg baths and manometers.

The Cambridge laboratory was a hive of activity—one of the world centres of modern biochemistry. The staff included a number of brilliant and enthusiastic biochemists. Among those who impressed me especially were Marjory Stephenson, Joseph and Dorothy Needham, Eric and Barbara Holmes, Malcolm Dixon, Robin Hill, and N. W. ('Bill') Pirie. I shared a laboratory with Bill Pirie, Kendall Dixon, and others. Among the Ph.D. students were David Green, Donald Woods, John Yudkin, Ernest Baldwin, and Leonard Stickland. There were close ties with neighbouring laboratories where research related to biochemistry was carried out by David Keilin, Geoffrey Roughton, Joseph Barcroft, Leslie Harris, Thomas Moore, and Alan Drury.

I continued the work I had begun at Freiburg on the metabolism of amino acids [48] and, as already mentioned in Chapter 6, I discovered that kidney, brain, and retina, as well as liver, readily synthesize glutamine from glutamate and ammonia, and that these tissues also contain two different glutaminases which can reverse the synthesis of glutamine [49]. I also spent much time on studying the fate of dicarboxylic acids in kidney tissue, a step in the work which led eventually to the concept of the citric acid cycle. With Norman Edson, a medical graduate and Ph.D. student from New Zealand, I studied the formation of ketone bodies in liver slices (1). I also started to follow up an observation made in Freiburg in joint work with T. H. Benzinger, which showed that in the pigeon the synthesis of uric acid—the main nitrogenous end-product of metabolism in birds—involves both liver and kidney [51, 52]. The great majority of birds form uric acid without participation of the kidney. This work led eventually, at Sheffield, to the discovery that in the pigeon a precursor of uric acid, hypoxanthine [65, 66], is formed. The kidney oxidizes it to uric acid.

It was made clear to me from the start that the supporting grant from the Rockefeller Foundation was for one year only, and in December Mr Miller of the Foundation advised me to accept any offer that might come my way. During the summer of 1933 I had received an attractive offer from Dr Joseph C. Aub of Harvard University, in whose laboratory at the Huntingdon Memorial Hospital I had worked for three weeks in 1929, and he repeated this offer in the spring of 1934. But I was disinclined to leave Cambridge as long as there was hope of staying there. Soon afterwards Hopkins told me that a post as demonstrator (the lowest step on the Cambridge academic ladder) might become vacant. He asked me whether I would be interested despite the low salary of £250 per annum. Without hesitation I said I would. He apologized for the inadequacy of the salary but told me that it seemed to be the only post which was likely to come up in the near future. I assured him that the low salary did not put me off, and he submitted the proposal through the academic channels. Some members of the University, though sympathetic to refugees, criticized Hopkins because they thought it wrong that a University post should go to a refugee at a time when there was unemployment among graduates in Britain. However, the opposition at the Board, if any, was not serious. But there was one more hurdle to cross. To be eligible for a University appointment it was necessary for me to become a member of the University through receiving the M.A. degree, conferred at an open meeting of the Senate. In case objections were raised, and to ensure a majority vote, Hopkins saw to it that the senior members of the laboratory attended the Senate on the day my degree was conferred, but there was no opposition of any kind.

The teaching duties associated with my new post began in the Michaelmas Term of 1934. Again there was some risk that students might protest and so Hopkins himself and a number of senior staff attended my first lecture. All went well.

Later in the term I introduced manometry to a class of 120

undergraduates. This was the first time that the Warburg manometer—then a novel and sophisticated instrument—had been used in large undergraduate classes. Up till then most class experiments had been simple test-tube work; manometry made it possible for students to watch the movements of the scale as cells took up oxygen or performed other complex processes. Before very long almost all major schools of biochemistry followed suit and adopted similar courses which introduced students to methods essential to the latest developments of 'dynamic' biochemistry and which demonstrated some of the key phenomena.

It was not long before I felt thoroughly at home in England. The 'British way of life' suited me down to the ground. For the first time I was living in a society virtually free from prejudice and permeated by a spirit of mutual respect and kindness. Germany even before the Nazis had been a place full of tension: party politics, religious antagonisms, anti-Semitism and the many social barriers put up by the military élite, the Prussian Junkers,[11] the exclusive student fraternities, all of which cut deeply into everyday life. People tended to be much more class-, race-, religion-, and party-conscious than in Britain, and their prejudices were reflected in their judgement and treatment of their fellow men.

In my day-to-day life in Cambridge the difference struck me in particular in people's helpfulness, their generous and warm hospitality and their innumerable kindnesses, large and small. I was invited to many homes. Vernon Booth, David Green, and Ann Barbara Callow helped me with my English and with writing scientific papers. Vernon later became an expert in this and his very successful booklet *Writing a scientific paper* (2) was published by the Biochemical Society and sold in thousands. He also taught me to drive in 1934 when I acquired a second-hand Austin 12 roadster for £22. When J. B. S. Haldane left his Cambridge Readership in Biochemistry in 1933 for a Chair of Genetics at University College, London, he offered his large Cambridge house free of charge to a group of refugees as a residence. We were all deeply touched and very grateful for this generous proposal and some of us were very much attracted, but the majority felt that it would be wrong for us to congregate under one roof.

One of the many personal human touches which impressed me was the handling of a maintenance grant to one of the refugee scientists in the Biochemical Laboratory, Dr Rudolf Lemberg. Lemberg, a biochemist from Heidelberg, had spent 1930–1 as a Rockefeller Travelling Fellow in the Laboratory. He was a very lovable person and when it became known in Cambridge in 1933 that he was in difficulty because of his Jewish origins, his friends and former colleagues collected a fund which enabled him to come to Cambridge. To prevent embarrassment the administration of the grant was placed in the hands of The Academic Assistance Council,[12] a London organization which collected and administered funds for the help of refugees. To Lemberg, then, the grant came from an anonymous body. He died in 1975 without ever knowing that the grant had come from friends and colleagues. I found this considerate and delicate handling of the situation very touching. The facts came to me confidentially at some time in Cambridge, but I think the lapse of time permits me to break the confidentiality.

With the help of a student of German literature with whom Eric Holmes put me in touch I set myself to improving my English. I also wanted to try to absorb some of Britain's literary heritage, so I began with nursery rhymes and Lewis Carroll and progressed to the classics, from Bacon and Addison to Wilde, Galsworthy, and Shaw. The latter three I had already read in translation. Shaw had long been my favourite contemporary writer but the German translations had been feeble and I enjoyed the original English tremendously.

The mastering of a new language, including its literary heritage, is a long-drawn-out process. After all, the education of a child takes many years. Although I already knew some English before I came to live here, it took perhaps ten years before I felt that I had a reasonably good command of the language. A book which I found most useful later—it appeared in 1939—was *This English language* by E. Denison Ross.[13] (Marrying in 1938 a Yorkshire girl with an exceptionally fine feeling for words and style also contributed much to the improvement in my knowledge and use of English.)

In the laboratory at Cambridge there was intense mutual interest in one another's research; ideas, difficulties, and results were openly and frankly discussed. There was also much light-hearted gaiety, wit, and humour. A record of this is the journal *Brighter Biochemistry* which was produced annually by the Laboratory from 1923 to 1931. It contained short, humorous stories, comic verse, drawings, and parodies of scientific papers contributed by members of the staff. By the time I arrived its publication had ceased but the spirit behind it was still much in evidence.

There was also a keen awareness of public affairs, national and international politics. From this sprang, among other things, the compassionate interest in refugees and their problems. Also keenly discussed were the relations between science and society. Early books on these subjects were *The social function of science* by J. D. Bernal (3), and *The social relations of science* by J. G. Crowther (4). Both had originally been Cambridge scientists. Though these books were written after my stay in Cambridge, it seems clear to me in retrospect that it was indeed no accident that they had a Cambridge background. The intellectual as well as the ethical climate there provided a hotbed where the concepts of the social responsibility of scientists could originate and grow. This happened at a time, it should be stressed, before the scale of the potential misuse and dangers of the new sciences—nuclear fission, destruction and pollution of the environment, biological warfare—had appeared on the horizon.

In 1934, Dr Redcliffe Salaman, whom I had come to know quite well, invited me to his home at Barley, twelve miles south of Cambridge, for Saturday 12 May, to meet Dr Chaim Weizmann and to discuss the possibility of my emigrating to Palestine (as it then was). One reason for my interest was the fact that it was often emphasized to refugees (myself included) that the generous hospitality extended to them in Britain was not necessarily permanent. Indeed my Home Office permit to remain in this country was issued on a yearly basis. In view of the extensive unemployment of the 1930s, public opinion was in two minds about playing host to people who might have to join the dole queue.

Dr Weizmann, who earlier in his career had been a Reader in Biochemistry at Manchester University, began by asking me to describe to him my work on the ornithine cycle of urea synthesis, and we then discussed what kind of possibilities there might be for me in Palestine. The idea emerged that a small group of refugee scientists might join one of the research centres, perhaps the Daniel Sieff Institute, the forerunner of the Weizmann Institute.

A potential member of the group to go to Palestine was David Nachmansohn, who was by this time working at the Sorbonne. He was a Zionist of long standing and he and Weizmann had known each other for many years. Early in 1936 Weizmann invited David and myself to make an exploratory visit to Palestine at his expense. This we did during the Easter vacation, travelling on a French ship from Marseilles to Jaffa. We visited various academic institutions, several kibbutzim, sometimes in the company of Dr Weizmann, and also went sight-seeing. It was a fascinating experience to see farming flourishing where there had been desert not many years before.

The research opportunities that could be offered turned out to be very limited. In the end, Dr Weizmann advised us that we should consider immigration only if we were satisfied that the facilities were adequate—after all, the essential object of our going there would be to carry out research.

I remained in touch with Dr Weizmann for some time afterwards, including the war years, but the opportunities for research which we had had in mind never materialized, neither for David Nachmansohn nor myself, and eventually we had struck roots too deep, in England and the States respectively.

The beauty of Cambridge and the East Anglian countryside, which I gradually explored, first on my bicycle and later by car, was a source of great pleasure to me, but by far the greatest satisfaction I derived was from the comradeship of the people around me.

I found very true what Carl Zuckmayer, himself a refugee from Nazism, once said: 'Home is not where a man is born, but where he wants to die, where he wants to carry out his life and bring it to a close as is ordained. It is where he has put his own roots down into the earth which he has broken by his own toil.'

In 1961 I had an opportunity of publicly relating my experiences and expressing my feelings about my stay in Hopkins's laboratory. This was when I was invited by the Biochemical Society to give the Third Hopkins Memorial Lecture

[182] in the Royal Institution on 28 March 1961. The major part of the lecture was scientific. At the end I tried to pay tribute to Hopkins. I described my correspondence with him and the events which led to my arrival in Cambridge, and I continued as follows:

The contrast between the world from which I had come away and the world which I entered was indeed enormous. Nazi philosophy had infiltrated all strata of German life and had created a dreadful atmosphere. It was a great shock to see how some of one's immediate colleagues, either because of conviction or because of blackmail and political pressure, adopted the official Nazi doctrines and cold-shouldered the political opponents of the Nazis. After this experience the warmth which greeted me at Cambridge from the minute of my arrival was indeed touching and heartening. There was a spirit of friendliness, of humanity, of tolerance and of fairness which had vanished in Germany. It was in Hopkins's laboratory where I saw for the first time at close quarters some of the characteristics of what is sometimes referred to as 'the British way of life'. The Cambridge laboratory included people of many different dispositions, convictions, and abilities. I saw them argue without quarrelling, quarrel without suspecting, suspect without abusing, criticize without vilifying or ridiculing, and praise without flattering. Hopkins was the central figure, beloved and respected as a natural leader, exercising leadership from within and not from above, utterly humble, modest, gentle, but by no means weak. For he had sufficient confidence in himself, as a leader must, to make decisions and to fight with tenacity for the aims in which he believed. Although President of the Royal Society, and 72 years old, he was accessible to all and always ready to help. I never heard him plead that he was too busy and had no time. His concern ranged far beyond biochemistry, Cambridge University and the Royal Society.[14] What struck me, in particular contrast with the German scene, was the strong 'social conscience' of Hopkins and his school, their deep concern for affairs of the world at large. One of its many manifestations was the practical help given, with substantial personal sacrifices, to refugees. Between 1933 and 1935 the laboratory sheltered six refugees from central Europe: Friedmann, Lemberg, Chain, Weil-Malherbe, Bach, and myself.

To me, as to many others, life in Hopkins's laboratory was a great education, a model of how true and lasting strength can be achieved in a team of scientists. Biological principles teach us that the strength of any population derives from the diversity of its individual members. A population is more likely to survive hazards the greater its diversity.

CAMBRIDGE (1933–1935)

Hopkins's school never concentrated on a few limited topics as many other schools did. His own research group was small. He encouraged the team to be versatile; everyone competent and enthusiastic was given the chance to venture ahead with his own ideas. This was probably a major cause of the far-reaching influence which the Cambridge school exerted on the development of the subject. It safeguarded the school against one of the worst hazards which can befall a team of scientists, the danger of losing the spirit of adventure and becoming obsolete in outlook and method.

You will appreciate that to me Hopkins is not only the great biochemist but also a man to whom I became deeply attached for very personal reasons. At a critical stage of my life, when the country of my birth proscribed me, he helped me, though a stranger, and welcomed me to the community of his laboratory, and together with his collaborators he initiated me into a new society. Sooner than I thought possible I was transformed from the stage of the visitor to that of a member. A token of this transformation was a touching set of incidents at the annual Sir William Dunn dinners, where, with Hopkins presiding, the memory of the benefactor of the Cambridge School of Biochemistry was honoured. When I attended this for the first time, after almost a year's stay, I was asked to reply to the Toast of the Visitors from Overseas, and a year later I was invited to propose, on behalf of the School, the Toast of the Visitors from Overseas. I cannot help feeling that the invitation of your Committee to stand in this place today, on the special occasion of the 100th Anniversary of Hopkins's birth and the 50th Anniversary of The Biochemical Society, is in the tradition of the very same spirit, demonstrating the acceptance of the one-time stranger into the family of British biochemists. For this I shall always be truly grateful.

In May 1935 the British Pharmacological Society met at Cambridge and I attended some of the sessions. During an interval I was approached by Professor Edward J. Wayne, who told me of a vacancy in his Department of Pharmacology at Sheffield University. Dr Douglas C. Harrison, who held the Lectureship in Pharmacology, had accepted the Chair of Biochemistry at Queens University, Belfast, and Wayne wanted to appoint another biochemist because there was no department of biochemistry at Sheffield. My name had been suggested to him, as I learned much later, by Charles Harington and George Pickering with whom Wayne had been associated at University College, London. The post attracted me because there seemed to be plenty of space, in contrast to Cambridge where the laboratory was very congested. Moreover, the appointment was a semi-permanent one and the salary £500 per annum, double that of the Cambridge post. We agreed that I should make an exploratory visit to Sheffield in September.

I was received in a most friendly way by the Waynes, with whom I stayed, by the Dean of the Medical Faculty, George Clark, and by Sir Arthur Hall who had retired from the Chair of Medicine but who was still influential as an elder statesman in the university. On my return to Cambridge I discussed the proposition with Hopkins who said he was sorry that he was not in a position to offer me a similar post. On weighing up all the circumstances I was influenced in favour of Sheffield by the consideration that there I would be able to expand my team effectively without trespassing on anyone else's claim. Although I was legally bound to give a term's notice at Cambridge, Hopkins released me on the understanding that I would continue my teaching responsibilities during the Michaelmas Term of 1935, which meant spending three weeks in Cambridge in November. To this Sheffield agreed, and my appointment as a Lecturer in Pharmacology was formally approved early in October 1935.

8
SETTLING AT SHEFFIELD
(1935)

I moved to Sheffield in the middle of October 1935 and was to spend nineteen happy years there, though at times world politics and the Second World War cast many dark shadows. I took with me the equipment which I had brought from Freiburg; a special grant of £100 by the University was to supplement it. Professor Wayne and his wife, Nan, were extremely kind and helpful. He gave me as much—or rather more—space in his laboratory than he could really afford and obtained additional funds for me for a technician. Wayne gave me invaluable advice on innumerable occasions, introduced me to many important aspects of British academic life, and he and his wife spent much time on improving my English. He went over the style and substance of my scientific papers with me, and as he had obtained a Ph.D. in chemistry before reading medicine, he was a highly competent judge and critic in my field. The Waynes also introduced me socially and asked me to dinner at their home nearly every week. Wayne's main professional interests were on the clinical side; he had beds and an out-patient clinic at the Sheffield Royal Infirmary. Although I was styled 'Lecturer in Pharmacology' he did not expect me to give regular lectures. However, on one occasion, when he was suddenly incapacitated by appendicitis, I was called upon—as Lecturer in Pharmacology —to take his place as a teacher. Luckily for me, the pharmacology of hormones and vitamins, a subject with which I was sufficiently familiar, fitted into the schedule. Considering that I was the only lecturer in his Department, so that Professor Wayne had to do all the teaching in pharmacology, the freedom he gave me was indeed generous. He encouraged me to lecture on biochemistry, a subject not officially represented in the University, and this was welcomed by the Professor of Chemistry, G. M. Bennett and the Professor of Physiology, George Clark. So I regularly gave two series of lectures, one to chemists reading for an Honours Degree and the other to students of the Honours School of Physiology.

A few weeks after my move, Hopkins wrote to say that a Lectureship had unexpectedly become vacant at Cambridge owing to the resignation of Dr T. S. Hele who had been elected Master of Emmanuel College, and he asked me whether I was interested. I replied that I was much attracted but Sheffield had made such a major effort to meet my wishes for equipment and space that I felt morally bound to stay for a while. In January 1936 he wrote again.

Dear Krebs,
The filling of our vacant Lectureship is still worrying me to death: I feel the responsibility very greatly, because the appointment will long outlast my own stay in the Department.

I want just to re-open the question of your own possible return; but only quite provisionally without committing you *in the least*.

Suppose we were to decide to get through this year without making the appointment and were able, next October, to offer you a post to be called 'Lecturer and Director of Research' with a salary about double that of a Lecturer, would you feel that the year's service you will then have given to Sheffield will justify you in leaving them?

It is a shame to disturb your mind again. Don't trouble to write more than just a word—just yes or no—if you feel that a decision is possible for you at this time.

I ought to say that the authorities have not given consent to the above suggestion, but I am told they might. I want at least to bring it before the Appointments Board if you think it at all possible for you.

This letter caused me much heart-searching. I showed it to Wayne and the Dean, who commented that if Cambridge would offer more than Sheffield, Sheffield would not stand in my way; they would consider it fair if I preferred Cambridge. But whether Cambridge in fact offered more was difficult to sort out. Money was not the only consideration: Sheffield offered better facilities in terms of space. On the other hand, to a biochemist the Cambridge environment was certainly much more stimulating and inspiring; in fact, there was no other biochemist in the Sheffield area. But I felt that I was not much in need of day-to-day stimulation and in any case Cambridge was within easy reach. A further point which arose in my discussions with Wayne and the Dean was the possibility of a

Department of Biochemistry being established at Sheffield, though it was impressed upon me that this was no more than a desirable aim. But it was my impression, confirmed by subsequent events, that this was not merely a matter of having a carrot dangled in front of me, but a genuine possibility.

Sheffield was a small university with around eight hundred students; the annual intake of medical students was only about fifty. Its smallness made it easy for close contacts to be maintained between departments and between faculties and these ties were further strengthened by regular social events. The amount of space at my disposal, coupled with the fact that I was very happy there and loved its beautiful surroundings with the Peak District touching the edge of the city, made me decide to stay. Sometimes I have wondered whether my decision would have been different had I been exposed to the temptations of college life during my two years at Cambridge. I had had no college affiliations while I had been there and had led a very quiet private life so that my decision was based essentially on the comparison of the opportunities for research. Even so, it was not easily reached, but never regretted because the next eighteen months were exceptionally fruitful in my research. They ended with the publication of the citric acid cycle concept [57, 58].

How the work on the citric acid cycle developed, and the importance of the contributions of other investigators in this field, is described in the next chapter. My most valuable helpers during this time were Leonard Eggleston and William Arthur Johnson. Eggleston joined me as a technician in January 1936 at the age of seventeen. He became a reliable, resourceful and long-standing collaborator, accompanying me to Oxford University and later into my 'retirement' post: he was with me until his sudden and untimely death in January 1974. The appointment of Johnson in April 1936 as a research assistant was made possible by a grant from the Medical Research Council. He had obtained a first-class degree in chemistry at Sheffield and was strongly recommended by the Head of the Department. The joint work on the tricarboxylic acid cycle was part of his Ph.D. thesis. Johnson learned biochemical methods very rapidly and was an excellent experimenter. Because of lack of funds I was unable to retain him after he had completed his Ph.D. work and he accepted a post in industry, with British Drug Houses.

He served in the Royal Air Force in the Second World War and soon after the war I lost touch with him. From time to time I made efforts to trace him but these were unsuccessful until as recently as 1978 when I located him in the Cayman Islands—now the manager of a successful turtle farm.[15]

When the evidence in support of the cycle seemed sufficient, I believed that the work might be of general interest and on 10 June 1937 I sent a 'letter' to the Editor of *Nature*. The paper was returned to me five days later accompanied by a letter of rejection written in the formal style of those days. This was the first time in my career, after having published more than fifty

```
Telegraphic Address:               Editorial and Publishing Offices:
PHUSIS, LESQUARE, LONDON.          MACMILLAN & CO., LTD.,
Telephone Number:                  ST. MARTIN'S STREET,
WHITEHALL 8831.                    LONDON, W.C.2.


RAG.AH/N.                          14th June 1937.


       The Editor of NATURE presents his compliments to
Mr. H. A. Krebs           and regrets that as he has
already sufficient letters to fill the correspondence
columns of NATURE for seven or eight weeks, it is
undesirable to accept further letters at the present
time on account of the delay which must occur in their
publication.
If   Mr. Krebs            does not mind such delay,
the Editor is prepared to keep the letter until the
congestion is relieved in the hope of making use of it.
He returns it now, however, in case  Mr. Krebs
prefers to submit it for early publication to another
periodical.
```

Fig. 4. Letter from *Nature* rejecting the first paper on the citric acid cycle, June 1937.

papers, that I experienced a rejection or semi-rejection. My 'letter' to *Nature* was never published. Instead, two weeks later, I sent the full paper to the journal *Enzymologia* in Holland and it was published [57] within two months. If one glances today at *Nature* of 1937 and examines the kind of papers which the Editor felt ought to be printed, one is struck by the difficulties that exist in deciding what is worth publishing. Most of the material published in the form of letters at that time has turned out to be of no particular importance.

It so happened that some sixteen years later, six days after the award of the Nobel Prize (for the work in which *Nature* had shown no interest) I received a letter from the Editors of that journal, again couched in the same formal language. They congratulated me on the award and asked me to review a paper for publication in the correspondence column. In my reply I could not refrain from adopting their style or from 'begging leave to bring to the notice of the Editors' the rejection of 1937. I received in return a gentle letter of apology and explanation. I believe that the story later became a cherished one in *Nature*'s editorial office.

During the middle 1930s the expansion of Hitler's powers and his threats to world peace became an increasing source of great anxiety. Like many of my fellow refugees who had first-hand experience of the monstrosities and absurdities of Nazism I had optimistically believed that the regime could not possibly survive for more than a few years. A collapse from within could not be expected because of Hitler's ruthless suppression of every potential opponent, but I believed that the self-interest of the rest of the world, particularly of Britain and France, would lead to an intervention. So it was tantalizing to watch one opportunity after another being missed until war became inevitable. Refugees were in a weak position to argue the case, or to participate in politics in any way. We were told to keep quiet and not misbehave by stirring up international trouble; we were thought to be emotional in our judgement because of our unhappy experiences at the hands of the Nazis. I was reminded of this situation recently when Solzhenitsyn in 1976, after warning the West of potential dangers from Russia, was told by newspaper editorials and letters to editors that a refugee with a deep grudge against his country of origin could not be an objective and fair judge of the international scene. In the late 1930s the warnings of far-seeing politicians, foremost among them Winston Churchill, were to us a bright light, but they were ignored. Sir Richard Acland recalls (1), 'We screamed our heads off trying to warn our countrymen about what was coming.' Sir Archibald Sinclair, then leader of the Liberal Party, pointed out in the House of Commons (referring to Job's lament that his adversary had not given him warning of his intentions) that the Prime Minister, Neville Chamberlain, had no such excuse for his foredoomed policies; it was evident that Hitler intended to carry through to the letter the policies and threats he had laid down in *Mein Kampf*. But too many politicians still believed that *Mein Kampf* was not a blueprint for action but mere propaganda. It was also worrying that Hitler had admirers and followers in this country, with Sir Oswald Mosley at their head. One day in 1938 I found on the windowsills of the staircase leading to my laboratory—where I would be sure to see them—hundreds of copies of an anti-refugee leaflet, which was printed on bright red paper (Fig. 5). Its general tone reminded me ominously of the posters I had seen in Germany. This was the only manifestation of this kind of personal hostility that I ever came across in Britain.

Such depressing experiences and thoughts did not interfere with my academic and personal affairs. War or no war, ordinary people carried on as usual. The authorities did not accept offers of direct help with the war effort but instructed people to carry on with 'business as usual' until orders were given to the contrary.

In 1938, at the initiative of Professor Wayne, my title of Lecturer in Pharmacology was changed to Lecturer in Biochemistry. This implied that there now was a Department of Biochemistry at Sheffield with myself at its head. I had a separate budget, but was still accommodated in the Department of Pharmacology. New premises were expected to become available in 1939 on the completion of a new building but because of the war our transfer was delayed until 1942.

In setting up the new and independent Department I received effective—and unasked for—help from the Rockefeller

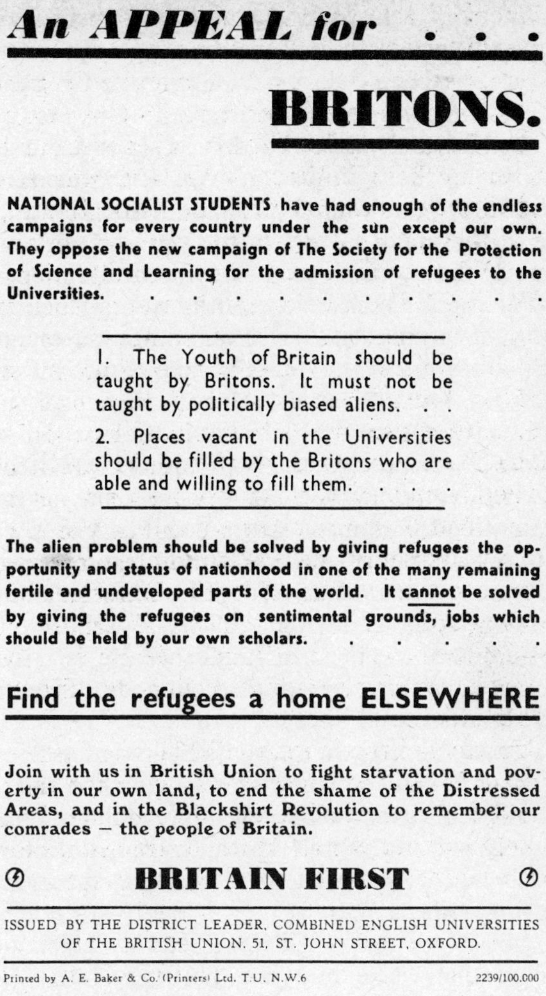

An APPEAL for . . .
BRITONS.

NATIONAL SOCIALIST STUDENTS have had enough of the endless campaigns for every country under the sun except our own. They oppose the new campaign of The Society for the Protection of Science and Learning for the admission of refugees to the Universities. . . . .

> 1. The Youth of Britain should be taught by Britons. It must not be taught by politically biased aliens. . .
>
> 2. Places vacant in the Universities should be filled by the Britons who are able and willing to fill them. . . .

The alien problem should be solved by giving refugees the opportunity and status of nationhood in one of the many remaining fertile and undeveloped parts of the world. It cannot be solved by giving the refugees on sentimental grounds, jobs which should be held by our own scholars. . . . . .

Find the refugees a home ELSEWHERE

Join with us in British Union to fight starvation and poverty in our own land, to end the shame of the Distressed Areas, and in the Blackshirt Revolution to remember our comrades — the people of Britain.

⚡ BRITAIN FIRST ⚡

ISSUED BY THE DISTRICT LEADER, COMBINED ENGLISH UNIVERSITIES OF THE BRITISH UNION. 51, ST. JOHN STREET, OXFORD.

Printed by A. E. Baker & Co. (Printers) Ltd. T.U. N.W.6   2239/100,000

Fig. 5. Anti-refugee leaflet circulated at Sheffield University, 1938.

Foundation. In 1936 the Foundation had started to support my research by an annual grant of £200. Early in 1938 their officers, Dr Tisdale and Dr Warren Weaver, asked me to meet them in London and informed me that the Trustees would prefer to consider a wider scheme of assistance, covering a period of five years. They stated that this would be conditional upon the establishment of a Department of Biochemistry. They were aware that this had already been proposed by Professor Wayne but they required confirmation from the Vice-Chancellor in the form of an official application from the University on these terms. The University agreed to do this. This generous support from the Rockefeller Foundation, which continued for a period of almost thirty years while I was in both Sheffield and Oxford, was one of the major factors in the provision of my research facilities. By the 1960s the grant was about £3,300 per annum.

In 1938 I attended the Seventeenth International Congress of Physiological Sciences at Zürich and on this occasion, for the first time since my emigration, I came across many German scientists who had been well known to me before 1933. They attended the Congress as a 'delegation' under a leader. All had been carefully screened by the Nazi authorities. Otto Warburg had independently notified the Congress organizers that he wished to attend, but at the last moment he was instructed by the Nazis to 'cancel participation without giving reasons'. With some measure of courage he sent a wire: 'Instructed to cancel participation without giving reasons. Warburg'. Professor A. V. Hill one of the most distinguished British participants, was keen to have the telegram read out at a full meeting of the Congress but its President, Professor W. R. Hess, pleaded that it might lead to unpleasant international repercussions and that the organizers ought not to risk annoying the rulers of Germany. So Warburg's message was not read out in public, but it was 'leaked' out and rapidly spread by word of mouth.

Some of the German participants seemed embarrassed in the company of refugees. Others were friendly but pretended that nothing had happened; when we raised the subject of politics, they asked that we should not discuss it because we were attending a scientific conference. Still others, notably Hans H. Weber, Professor of Physiology at Münster University and later at Königsberg, with deliberate ostentation associated with the refugees and wanted to be seen in their company.

At this time I was still a German subject; one could not apply for naturalization until after five years' residence in Britain. In 1938 I had to apply for a renewal of my passport and the German Consul at Liverpool informed me that he would be willing to do this but,

SETTLING AT SHEFFIELD (1935)

according to a communication from the authorities in Germany, your entry into Germany, even for a temporary stay, is undesirable.

When the passport was issued four weeks later, the Consul repeated the warning and added

should you nevertheless return to Germany without permission you would have to reckon with the measures provided for emigrants.

In the spring of 1938 I married Margaret Fieldhouse, the daughter of a Yorkshire family and a teacher in domestic science at Sheffield. We had been introduced by a mutual friend, the botanist John Lund, who was at that time a temporary Assistant Lecturer. A year after our marriage our first child, Paul, was born and was followed at three-yearly intervals by Helen and John.

In 1938 also the Department's first post-doctoral visitors from overseas, Dr Åke Öström from Sweden (a Rockefeller Travelling Fellow) and Philip P. Cohen from the University of Wisconsin (a US National Research Council Fellow) arrived to contribute effectively to the work of the laboratory [65, 66, 67, 69, 70, 73]. In May 1939 Earl E. Evans, Jr. from Chicago, also a Rockefeller Travelling Fellow, joined us [77, 78].

Apart from working on several aspects of the citric acid cycle, the laboratory was concerned with the synthesis of purines and uric acid (referred to in Chapter 4) [65, 66], the synthesis of glutamine [67] and α-oxoglutarate [77] in pigeon liver (Chapter 11) and the study of transaminases [70]. The observation that pigeon liver readily converts pyruvate into glutamine and α-oxoglutarate, two closely related substances:

$$COOH.CH_2.CH_2.CO.COOH \quad \text{α-oxoglutarate}$$
$$CONH_2.CH_2.CH_2.CHNH_2.COOH \quad \text{glutamine}$$

proved to be an important step in the discovery of CO_2 fixation in higher animals (see Chapter 11). The experiments on transaminases by Philip Cohen were started after Braunstein and Kritzmann (2) had reported the discovery of the existence of transaminating enzymes. Braunstein had limited himself to essentially qualitative aspects because of the lack of convenient quantitative methods. Cohen first worked out a quantitative method for the determination of glutamate [69] and found that the transaminase of glutamate oxaloacetate is far more active than any other [70].

In June 1938 I applied for naturalization. Proceedings took about a year and in August 1939 I was informed that the papers would be granted. Unfortunately they were not completed until 6 September 1939. In consequence I was an 'enemy alien' on 3 September, the first day of the war, and my car was confiscated. However, it was returned three days later when the naturalization procedures had been completed.

I was, of course, anxious to offer my services to the war effort, but they were not wanted. Once again I was told to 'carry on as usual'; to keep my university work going would be the best contribution I could make.

A few days after the outbreak of the war I received news from relations living in Luxemburg that my father had died suddenly from a stroke. The message came as a tremendous shock. He had been in good health when he had last visited me in August 1937 and had remained so until his last day. He was then 72. During the 1930s my sister and her husband had emigrated to Israel and my brother had come to Britain. My father had remained in Hildesheim. Later I realized that his death at this time had spared him the fate of many of his generation who were to be exterminated in the ensuing years. Nazi persecution had made his last years very sad. He had been deprived of the right to practise, much of the money he had saved for old age had been extorted by 'special levies', and at the time of the pogrom in November 1938 he had been rounded up and imprisoned for about a fortnight. Early in 1939 we managed to arrange for my young sister, Gisela, to become a pupil at a convent school in Leeds, which also accepted her mother as a teacher. During the summer vacation of 1939 they went home to visit my father and the outbreak of war prevented their return.

After the German invasion of Western Europe in the spring of 1940 until the end of the war, no communication of any kind was possible with my relatives and friends in Germany.

9
THE HISTORY OF THE DISCOVERY OF THE CITRIC ACID CYCLE (1936–1937)

In this chapter I will make an attempt to describe for the reader with minimal knowledge of chemistry the work which led to the discovery of the citric acid cycle and the physiological role of the cycle. Those with no grounding in chemistry may miss out this chapter.

Some people who have read about the cycle in textbooks have told me that they cannot imagine how I conceived such a complex sequence of reactions. Indeed the concept was not the result of a sudden inspiration but of a very slow evolutionary process extending over some five years and beginning, as far as I was concerned, in 1932. As I will explain, many investigators contributed towards the elucidation of the cycle by making discoveries which by themselves were not sufficient to lead to the concept of the cycle, but which nevertheless provided important pieces of information. Eventually these proved to be essential building blocks in the construction of the cyclic concept.

The citric acid cycle defines some of the chemical changes which foodstuffs—carbohydrates, fat, protein—undergo when they are burned, i.e. oxidized. This combustion is one of the fundamental chemical reactions in living organisms because it provides energy, and without a continuous supply of energy life cannot exist. Energy is needed for building up and maintaining living structures and for every activity associated with life, such as movement, growth, secretion, and the synthesis of body cell materials, including hormones and enzymes.

Earlier work

At the time when I began to work in this field, the 1930s, there was very little information about the finer details of biological combustion. We knew from Lavoisier (1789) that the changes which foodstuffs undergo in the body are combustions, and that the end-products are mainly carbon dioxide and water. Thus we knew the 'overall' process. When glucose—the main carbohydrate—is oxidized, the overall reaction is expressed by the equation

$$\underset{\text{Glucose}}{C_6H_{12}O_6} + \underset{\substack{\text{Oygen} \\ \text{gas}}}{6\ O_2} \longrightarrow \underset{\substack{\text{Carbon} \\ \text{dioxide}}}{6\ CO_2} + 6\ H_2O$$

The principles of chemistry teach us that the six oxygen molecules cannot all react at the same time because, as a general rule, an individual chemical process is the reaction between two substances only. This implies that the combustion of glucose is a process involving many separate steps which require identification.

By 1932 it was known that glucose is not burned until it has first been 'fermented' (that is, a fission of the glucose molecule not involving oxygen of the air). Fermentations of sugars represent a minor source of energy. Thus, a muscle can contract for a while to do work (e.g. to lift a weight) even when the oxygen supply has been stopped by the application of a tourniquet. Yeast grows in the absence of oxygen during the manufacture of beer and wine. Here the source of energy is a 'fermentation'. In muscle, glucose is split into two molecules of lactate[16]

$$\underset{\text{Glucose}}{C_6H_{12}O_6} \rightarrow \underset{\text{Lactate}}{2\ C_3H_6O_3.}$$

In yeast the fission leads to the formation of alcohol and CO_2

$$\underset{\text{Glucose}}{C_6H_{12}O_6} \rightarrow \underset{\text{Alcohol}}{2\ C_2H_5OH} + 2\ CO_2$$

The fermentations in muscle and yeast are chemically very similar although the end-products are different. Glucose and the products of fermentation contain the same numbers of C, H, and O atoms, but these are arranged in different ways, as shown in the following formulae

DISCOVERY OF THE CITRIC ACID CYCLE (1936–1937)

$$
\begin{array}{ccc}
\text{HCO} & & \\
| & & \\
\text{HCOH} & \text{CH}_3 & \text{CH}_3 \\
| & | & | \\
\text{HOCH} & \text{CHOH} & \text{CH}_2\text{OH} \\
| & | & \\
\text{HCOH} & \text{COOH} & + \\
| & & \\
\text{HCOH} & & \text{CO}_2 \\
| & & \\
\text{CH}_2\text{OH} & & \\
\\
\text{Glucose} & \text{Lactic} & \text{Alcohol} \\
\text{(simplified} & \text{acid} & + \\
\text{formula)} & & \text{CO}_2
\end{array}
$$

In glucose, most of the carbon atoms have attached to them one H and one OH. In lactic acid, one carbon has lost this O to become CH_3, another has gained one O to become COOH. In the course of this rearrangement, energy is set free. The structure of lactic acid is related to that of alcohol: alcohol can be looked upon as lactic acid which has lost CO_2:

$$
\begin{array}{cc}
\text{CH}_3 & \text{CH}_3 \\
| & | \\
\text{CHOH} \longrightarrow & \text{CH}_2\text{OH} \\
| & + \\
\text{COO}\!:\!\text{H} & \text{CO}_2
\end{array}
$$

However, the detachment of CO_2 from the other two carbon atoms occurs before the formation of lactic acid. The immediate precursor of lactic acid is pyruvic acid which in animal tissue takes up two H atoms to form lactate, but in yeast loses CO_2; it is the remaining compound, acetaldehyde, which takes up the two H atoms to form alcohol:

$$CH_3.CO.COOH \text{ (Pyruvic acid)}$$

Animal tissue ↙ +2H   Yeast ↘

$$CH_3.CHOH.COOH + CO_2$$ $$CH_3.COH \text{ (acetaldehyde)} + CO_2$$

↓ +2H

$$CH_3.CH_2OH \text{ (alcohol)}$$

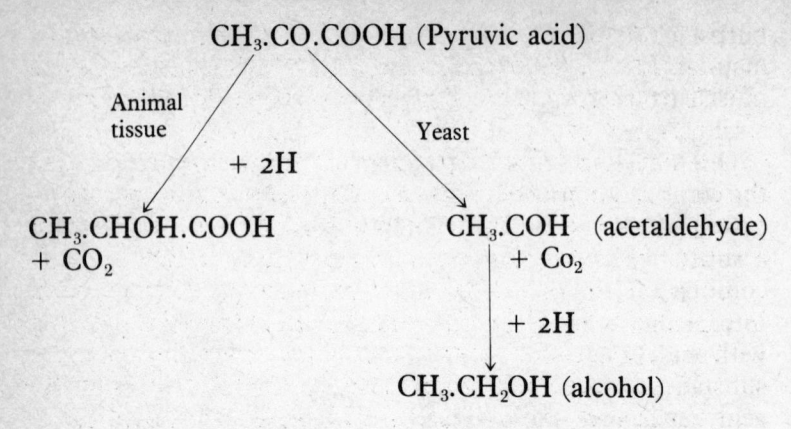

What is burned then, is the product of fermentation, i.e. lactate and pyruvate in animal tissue, alcohol and pyruvate in yeast. Like fermentation, combustion has many stages. Both lactic acid production and alcohol production are the result of about twelve different steps.

But the individual steps of combustion were unknown. There were technical difficulties to be overcome before they could be studied. The standard methods of chemistry permit a study of chemical reactions only when the reactions can be made to occur in solution and so only after biochemists had worked out methods of bringing the lactic acid formation of muscle and the alcoholic fermentation of yeast into solution, did it become possible to study the individual steps and to identify the intermediate substances. By contrast the combustion processes of muscle and yeast or any other cell proved to be closely tied to insoluble cell structures. When attempts were made to extract the oxidative processes by grinding up tissues with aqueous solutions, the processes no longer occurred. It was thus necessary to develop tissue preparations in which the energy-supplying structures remained intact.

The contributions of Albert Szent-Györgyi

The Hungarian-American biochemist Albert Szent-Györgyi developed a tissue preparation in which the key structure for oxidations—the mitochondrion—stayed intact. He used the flight muscle of pigeons, an exceedingly active material which burns foodstuffs at a very high rate (for flight can only be maintained by a very rapid energy supply). He coarsely minced muscle from a freshly killed pigeon and suspended the mince in a saline medium.

One way of tackling the problem of the intermediate steps of the combustion process is to test which substances, apart from carbohydrate and fat, burn the most readily. The logic is that if a substance is an intermediate then it must readily undergo combustion: if it proves to be non-combustible, it cannot be an intermediate. Szent-Györgyi and others before him working with somewhat similar methods noted that among innumerable substances tested, only a few, namely succinic acid, fumaric acid, malic acid, and oxaloacetic acid, were readily oxidized in suspensions of minced muscle. The formulae of these substances show that all four are very similar. All are 4-carbon dicarboxylic acids and differ only in respect of the hydrogen and oxygen atoms attached to the carbon skeleton. Succinic acid can, in fact, be converted into the other three, as indicated by the arrows:

(A)

$$\begin{array}{c}\text{COOH}\\|\\\text{CH}_2\\|\\\text{CH}_2\\|\\\text{COOH}\end{array} \xrightarrow{-2H} \begin{array}{c}\text{COOH}\\|\\\text{CH}\\\|\\\text{CH}\\|\\\text{COOH}\end{array} \xrightarrow{+H_2O} \begin{array}{c}\text{COOH}\\|\\\text{CHOH}\\|\\\text{CH}_2\\|\\\text{COOH}\end{array} \xrightarrow{-2H} \begin{array}{c}\text{COOH}\\|\\\text{CO}\\|\\\text{CH}_2\\|\\\text{COOH}\end{array}$$

Succinic acid   Fumaric acid   Malic acid   Oxaloacetic acid

However, the knowledge that these substances are readily oxidized did not, to begin with, throw any light on the oxidation of foodstuffs, because their structure seemed to have no connection with those of foodstuffs. Nor was it clear what substances might be intermediates between oxaloacetate and the end-products, carbon dioxide and water.

Szent-Györgyi made the most important discovery: he showed that the rate of oxidation in minced muscle suspension can be greatly increased by adding very small quantities of any of these dicarboxylic acids. When he measured the amounts of extra oxygen taken up after addition of these substances, he found that the extra oxygen consumption could not be explained merely as an oxidation of the added substances. Thus he rightly concluded that these substances can 'catalytically' accelerate the combustion of substances contained in the muscle suspension. But he could not offer a satisfactory explanation for this catalytic effect.

The contributions of Knoop and Martius

The next step in providing relevant information was a discovery by the German biochemists, Knoop and Martius. These investigators did not set out to study the combustion of foodstuffs; their object was to establish the intermediate stages of the oxidation of citric acid. Small amounts of citric acid are present in many foodstuffs. They can serve as a source of energy by undergoing combustion but here again, as in the case of the main foodstuffs, the chemical steps which convert citrate to carbon dioxide and water were obscure. Knoop and Martius clarified the initial stages of the combustion. They discovered the following sequence of reactions in liver tissue:

(B)

$$\begin{array}{c}\text{COOH}\\|\\\text{C}\begin{array}{c}\diagup\text{OH}\\\diagdown\text{CH}_2.\text{COOH}\end{array}\\|\\\text{CH}_2\\|\\\text{COOH}\end{array} \xrightarrow{-H_2O} \begin{array}{c}\text{COOH}\\|\\\text{C}.\text{CH}_2.\text{COOH}\\\|\\\text{CH}\\|\\\text{COOH}\end{array} \xrightarrow{+H_2O}$$

Citric acid   Aconitic acid

$$
\begin{array}{l}
\text{COOH} \\
| \\
\text{C}\!\!<\!\!\begin{array}{l}\text{H}\\ \text{CH}_2.\text{COOH}\end{array} \\
| \\
\text{CHOH} \\
| \\
\text{COOH}
\end{array}
\quad \xrightarrow{-2\text{H}} \quad
\begin{array}{l}
\text{CO}_2 \\
+ \\
\text{CH}_2.\text{CH}_2.\text{COOH} \\
| \\
\text{CO} \\
| \\
\text{COOH}
\end{array}
$$

Isocitric acid α-Oxoglutaric acid

The end-product of these sequences, α-oxoglutaric acid, was already known as an intermediate product because it is formed from the amino acid glutamate and had been shown to be oxidized to succinic acid:

$$(C) \quad \text{COOH.CH}_2.\text{CH}_2.\text{CO.COOH}$$
α-Oxoglutaric acid

$$\downarrow +\text{O}$$

$$\text{COOH.CH}_2.\text{CH}_2.\text{COOH} + \text{CO}_2$$
Succinic acid

The discoveries of the sequences (A), (B), and (C) in 1937 established that there was a set of reactions in the liver which converts citrate to oxaloacetate:

Citrate → aconitate → isocitrate → α-ketoglutarate → succinate → fumarate → malate → oxaloacetate.

But it still was not clear what this sequence had to do with the combustion of the main foodstuffs.

The crucial experiments

This was the stage at which my own constructive contribution began. For some time, from 1932, I had tested the oxidizability in various tissues (especially kidney, liver, and muscle) of substances which, on the basis of knowledge of chemistry, might possibly be intermediates in the combustion of foodstuffs, and I had seen the ready oxidation of citrate. I, too, had tried to elucidate the chemical reactions of citric acid, but without success. My interest in this tricarboxylic acid and the dicarboxylic acids mentioned earlier stemmed from my conviction that their ready oxidation was most likely to be connected with the combustion of foodstuffs. I had three reasons for thinking this. First they were the *only* substances among many dozens examined which burned at about the same rate as foodstuffs. Secondly, almost all the properties of living matter have a function, i.e. if a substance or a process occurs it is likely to have a role to play in the life of the cell. This follows from the principle that in the course of evolution, non-functional properties do not, in general, survive. Thirdly it was known from earlier work reported by Thunberg (1) in 1910 and by Quastel (2) in 1928 that malonic acid,

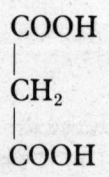

$$\begin{array}{c} COOH \\ | \\ CH_2 \\ | \\ COOH \end{array}$$

a substance similar in structure to succinate, specifically inhibits the oxidation of succinate to fumarate and the whole process of combustion in living cells. This indicated that the oxidation of succinate to fumarate is a component reaction of biological combustion. These considerations pointed strongly to a link between the oxidation of foodstuffs and the sequence of reactions which leads from citric acid to oxaloacetic acid. So I asked myself whether perhaps oxaloacetic acid, together with a substance derived from foodstuffs, might combine to form citrate again, after the manner of a cycle. On the basis of the biochemical information already available, a most likely candidate to react with oxaloacetate was pyruvate, which I have already mentioned as an intermediate in the anaerobic degradation of carbohydrate. So I used suspensions of minced pigeon flight muscle to test whether oxaloacetate and pyruvate together form citrate, and found that they did. The following reaction was established:

DISCOVERY OF THE CITRIC ACID CYCLE (1936–1937)

$$\begin{array}{c}\text{COOH}\\|\\\text{CO}\\|\\\text{CH}_2\\|\\\text{COOH}\end{array} \xrightarrow{\text{(Pyruvic acid)}\\ + \text{CH}_3.\text{CO.COOH}} \begin{array}{c}\text{COOH}\\|\\\text{C}{\diagup}^{\text{OH}}_{\diagdown \text{CH}_2.\text{COOH}}\\|\\\text{CH}_2\\|\\\text{COOH}\end{array}$$

Oxaloacetic acid　　　　　　　　　Citric acid

The finer details of the reaction remained unknown until 1951, seventeen years after the discovery of the process (see below). At this stage, however, the information was only qualitative. What remained to be demonstrated was that the rates at which the key reactions of the cycle occurred, especially the rate of synthesis and degradation of citric acid, were sufficiently high for the whole of the tissue combustion to pass through the stage of citrate. If the rate of the combustion process is known then the speed of the synthesis and the degradation of citrate can be calculated. Knoop and Martius had not carried out any quantitative studies, and had limited their studies to liver. I did the necessary measurements and found that in muscle tissue as well as in many other animal tissues the rates were sufficiently rapid to support the concept that the tri- and dicarboxylic acids did play a key role in the combustion of foodstuffs. This information made it possible to formulate a cyclic process in which the di- and tricarboxylic acids are intermediary stages. One turn of the cycle would bring about the oxidation of an equivalent of acetic acid $C_2H_4O_2$ as shown in Fig. 6.

There remained the question of how oxaloacetate and pyruvate react to form citrate. This could not be answered until fourteen years later after Lipmann showed that pyruvate first reacts with a coenzyme he had discovered, called coenzyme A, to form acetyl coenzyme A according to the equation

$CH_3.CO.COOH$ + coenzyme A + $\tfrac{1}{2}O_2$ → $CH_3.CO.CoA + CO_2$
Pyruvic acid　　　　　　　　　　　　Acetyl-CoA

It is the acetyl-CoA which reacts with oxaloacetate to form citrate and free CoA (3, 4):

$$\begin{array}{c}\text{COOH}\\|\\\text{CO}\\|\\\text{CH}_2\\|\\\text{COOH}\end{array} \xrightarrow[\text{+ CH}_3\text{.CO.CoA + H}_2\text{O}]{\text{acetyl-CoA}} \begin{array}{c}\text{COOH}\\|\\\text{C}{<}{}^{\text{OH}}_{\text{CH}_2\text{.COOH}}\\|\\\text{CH}_2\\|\\\text{COOH}\end{array} + \text{CoA}$$

Oxaloacetic acid   Citric acid

Thus CoA can react again and again to form acetyl-CoA.

I had first thought of carbohydrate as the main energy source of muscle tissue, for the simple reason that it was well established that muscle tissue can produce much lactic acid during exercise. Later, Lynen in Germany showed that fatty acids also supply acetyl coenzyme A. Gradually it became clear that many of the carbon atoms (the fission products of protein) are converted into acetyl coenzyme A. Thus all three major constituents of food supply carbon atoms in the form of the acetic acid attached to coenzyme A for combustion. The citric acid cycle forms two-thirds of the total combustion process in living cells; the remaining third is a process (also a combustion) by which carbohydrate, fatty acids, and amino acids are prepared for entry into the cycle. It is obvious that the cycle plays a central role in the supply of energy. Figure 6 illustrates this situation diagrammatically.

The cycle explains why the di- and tricarboxylic acids are readily oxidized in living matter. It also explains why the addition of small quantities of dicarboxylic acids to tissue preparation catalytically increases the rate of combustion—because these acids are regenerated in the course of the cycle.

Universal occurrence of the cycle

The cycle has been found to occur in every kind of animal and plant, down to the most primitive bacterium. This implies that it arose very early in evolution. Indeed the occurrence of the components of the cycle in plants accounts for some of their names, which may seem odd to the uninitiated. The names tell us the material in which the acids were first discovered: malic acid in apple (*Malus*), citric acid in citrus fruits, aconitic acid in

[Diagram of the citric acid cycle showing: Acetyl-CoA (formed from Carbohydrate, Fat, Protein) + H₂O → Citrate (releasing CoA) → (H₂O) → Aconitate + H₂O → Isocitrate → (2 H, CO₂) → α-Oxoglutarate + H₂O → (2 H, CO₂) → Succinate → (2 H) → Fumarate + H₂O → Malate → (2 H) → Oxaloacetate → back to Acetyl-CoA]

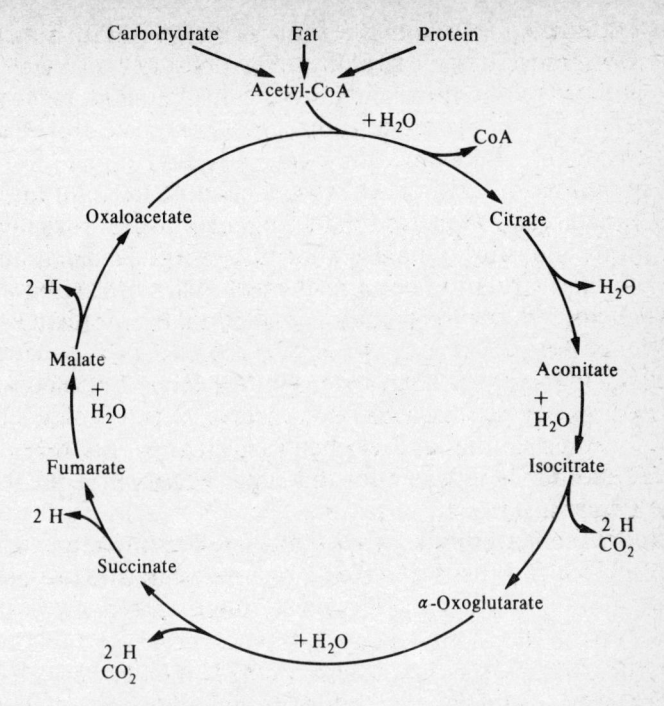

Fig. 6. Diagram of the citric acid cycle and the chemical processes leading from foodstuffs to the cycle. Carbohydrate, fat, and protein all form acetic acid attached to coenzyme A (CoA). CoA is regenerated when citrate is formed—thus it acts as a catalyst and only small amounts are therefore needed. Products formed by the cycle are written outside, the substances entering, inside. In the course of one turn of the cycle, as written, two CO_2 and four pairs of H atoms are formed. The latter react with O_2 to form H_2O. On balance, one turn of the cycle causes the combustion of one molecule of acetic acid

$$C_2H_4O_2 + 2O_2 \rightarrow 2CO_2 + 2H_2O$$

Apart from forming acetyl-CoA, amino acids also form α-oxoglutarate, pyruvate, and oxaloacetate; therefore part of the carbon skeleton of protein can join the pathway of degradation at stages other than at the synthesis of citrate. Much later one further intermediate was identified between α-oxoglutarate and succinate-succinyl-coenzyme A.

the monkshood (*Aconitum*), fumaric acid in the fumatory (*Fumaria*), succinic acid in amber (*Succinum*, a fossil resin from an extinct pine). In these and other plants the acids often occur in relatively high concentrations because, apart from being intermediate metabolites, they have a second function. They are present as the salts of calcium, potassium, magnesium, and other metals and as such store the minerals needed for growth. Isocitrate and α-oxoglutarate were given chemical names referring to their structure, because chemists had synthesized them long before their ubiquitous presence in living matter was known.

I should mention that in arriving at the concept of this cycle I was guided by my discovery five years earlier of the ornithine cycle of urea synthesis, described in Chapter 5. My mind was thus conditioned to watch for this type of reaction sequence in the living world.

Subsequent studies showed that there are numerous analogous cycles and the two cycles therefore established a general principle of the chemical organization of living matter.

For some time, until the early 1950s, there was doubt as to whether the citric acid cycle also occurred in micro-organisms, such as yeast and *Bacillus coli*. These doubts arose from the experimental observation that added citrate is either not utilized by these organisms, or is used exceedingly slowly. Thus a key experiment to establish the occurrence of the cycle in animal tissue could not be repeated with these micro-organisms. However, in the course of the 1950s we began to realize that non-utilization may be a consequence of non-penetration of citrate into the cell. Up until then we had tacitly assumed that small molecules readily pass through cell membranes, but this was a mistake. We now know that in many cases a complex process involving specific carrier proteins regulates the entrance of substances into cells. Yeast and *Bacillus coli* happen to lack such a specific carrier system for citrate. These carrier systems see to it that the internal environment of the cell is compatible with the life processes. Later, when isotopically labelled substances became available, we learned that citrate can be formed within the cell and readily undergo the reactions of the citric acid cycle.

Citric acid cycle or tricarboxylic acid cycle?

The cycle is referred to in the literature by three different names: 'citric acid cycle', 'tricarboxylic acid cycle', and 'Krebs cycle'. Many people prefer the last name, presumably because it is the shortest. I named it the citric acid cycle after its characteristic intermediate. Four years after the baptism, in 1941, evidence came to light which cast doubt upon the sequence in which the tricarboxylic acids arise. Experimentally the three acids always appear together and there was no information about the order in which they are formed from oxaloacetate and pyruvate. The formulation

$$\text{citrate} \rightarrow \text{aconitate} \rightarrow \text{isocitrate}$$

was therefore somewhat arbitrary. Then, in 1941, Harland Wood, and his colleagues at Ames, Iowa and Earl Evans at Chicago reported experiments carried out with radioactive carbon which were believed to indicate that citrate could not be a direct intermediate, and it was suggested that aconitate was the first tricarboxylic acid to arise from oxaloacetate and pyruvate. This was assumed to form isocitrate. Citrate was believed to be formed by a side reaction

$$\begin{array}{c} \text{oxaloacetate} \\ + \\ \text{pyruvate} \end{array} \rightarrow \begin{array}{c} \text{aconitate} \\ \updownarrow \\ \text{citrate} \end{array} \rightarrow \text{isocitrate}$$

These assumptions were generally accepted and I therefore renamed the cycle 'tricarboxylic acid cycle'. However, in 1948, Alexander Ogston showed, by a penetrating theoretical analysis, that the earlier doubts had been unfounded, and three years later Ochoa directly demonstrated that it is citrate which is formed from oxaloacetate and acetyl coenzyme A, and not aconitate or isocitrate. A full discussion of these developments is in Chapter 14.

Concluding remarks

This account of the historical development should make it clear that the pieces of information on which the concept of the cycle rests became gradually available through the efforts of several investigators. Eventually I could put them together like the pieces of a jigsaw puzzle. The main pieces, to summarize, were

1. The ready oxidizability of the di- and tricarboxylic acids in respiring tissues, and the absence of other readily oxidizable substances.
2. The conviction, based on evolutionary principles, that the oxidizability of these substances must be related to the combustion of foodstuffs.
3. The catalytic acceleration of combustion in tissue preparations by dicarboxylic acids.
4. The inhibition of oxidation in biological material by malonate, a specific inhibitor of the enzyme responsible for the oxidation of succinate. This implies that succinate must be an intermediate in cellular oxidation.
5. The demonstration that the degradation of citric acid leads to the formation of the dicarboxylic acids and that citrate can be synthesized from oxaloacetate and pyruvate, a product of carbohydrate degradation and that the rates of these reactions were high.

The first four were originally isolated pieces of information, the fifth made it possible to formulate a coherent story connecting all the facts. As already stated, in visualizing the cyclic mechanism it was of major relevance that five years earlier I had been concerned with the first metabolic cycle to be discovered, the ornithine cycle of urea synthesis.

So my main contribution was the discovery of one crucial fact—the synthesis of citrate from oxaloacetate and pyruvate in biological material—which provided the links between all the other observations concerning oxidations in biological material.

10
THE WAR YEARS
(1939–1945)

During the early stages of the war (the 'phoney war' from September 1939 until the German attack in the West in April 1940) life in the laboratory went on as before. I was more anxious than ever to contribute effectively to the war effort but no opportunity to bring research to bear on some practical wartime problem arose until 1941. Dr Kenneth Mellanby, a Research Fellow in the Zoology Department at Sheffield, had assembled a team of some twenty conscientious objectors from pacifist organizations who had offered to act as volunteers in experiments which, in a peaceful way, would benefit humanity. They were housed together at the Sorby Research Institute, so named because at that time Mellanby was Sorby Fellow of the Royal Society. The objectors came from many walks of life and were willing to take risks, making it clear that they had not evaded war service because of its unpleasantness and danger. Mellanby has described this work in *Human guinea pigs* (1), published in 1945 and republished with additional comments in 1973.

The first experiments designed by Mellanby, after extensive consultations with the Ministry of Health, were concerned with the parasitic skin disease, scabies, the main aim being to find out how the parasite is transmitted from person to person. In the course of these experiments Mellanby came to realize that he was not making full use of the opportunities that the volunteers offered—a group of intelligent and co-operative people living under closely controlled conditions was ideal for experimental work. He discussed the possibility of some nutritional experiments with them and they were quite enthusiastic about combining some such work with the scabies investigation. He approached me about collaboration and I agreed, but we thought that the decision on what kind of work should have priority should be made by higher authorities, perhaps by the

Government Food Policy Committee, the Ministry of Health, or the Medical Research Council. We were referred to Dr R. A. McCance and Elsie M. Widdowson at Cambridge, who were already concerned with nutritional experiments on human subjects. On their advice it was decided to study high extraction (85 per cent) wheatmeal (later called 'national wheatmeal') and to compare it, especially from the point of view of digestibility and its effects on calcium absorption, with a 75 per cent extraction flour [85]. Wheat was a staple in the British diet and a great deal had to be imported under armed convoy. The Government decision to make maximum use of wheat by higher extraction was prompted by the German attempt to starve Britain into surrender through U-boat warfare, but there were doubts that a higher level of extraction would make an appreciable contribution to the nation's food supply. One of the experts, Norman Wright, expressed the view that much of the extra flour obtained in this way could not be digested and would therefore be lost; if the extra flour were to be fed instead to dairy cattle, Wright thought, some of it would be converted into milk.

The volunteers were given diets containing 500 and 700 g of either high extraction wheatmeal or white flour, and the dry weight and nitrogen content of the faeces were measured. The results did not substantiate Wright's views: a high proportion of the extra meal extracted proved to be digestible and its nitrogen largely utilized.

A second problem for investigation suggested by Dr McCance was the effect of national wheatmeal on the absorption of calcium from the intestinal tract [94]. High extraction flours contain phytic acid, which forms insoluble salts with calcium and may therefore prevent the absorption of calcium. McCance and Widdowson had already demonstrated that this applied to 92 per cent extraction flours and our experiments confirmed it for 85 per cent extraction. These findings supported the official policy of adding small amounts of calcium salts to the 85 per cent extraction flour to compensate for diminished absorption.

A further opportunity for experiments with the volunteers came when the Medical Research Council's 'Accessory Food Factors Committee' asked for information on human vitamin requirements, in the first instance vitamin A [98, 121]. Reliable information was scanty. So a team of investigators was set up by the Vitamin A sub-Committee of the Medical Research Council to study this subject at the Sorby Research Institute. Twenty-three men and women volunteered to live on a diet deficient in vitamin A but complete in every other respect. The aim was to look for early signs of deficiency and to establish the dose which would prevent it. The experiment lasted a long time, from July 1942 until October 1944, because signs of deficiency (as measured by the vitamin A depletion of the blood and deterioration of dark-adaptation) took over a year to develop.

The Committee met in Sheffield at regular intervals. Its chairman was Dr R. A. Morton of Liverpool and its secretary Miss E. M. Hume of the Lister Institute. I was given responsibility for the day-to-day management of the volunteers when Mellanby left in late 1943 to join the army as a specialist in biological research.

The main conclusion we reached from these dietary experiments was that the body's reserves of vitamin A are normally sufficient to prevent the onset of definite signs of deficiency for twelve to twenty months (provided that subjects have previously been on a normal diet), and that a daily dose of 1300 International Units of vitamin A was sufficient to cure mild vitamin A deficiency.

This experiment was followed by a similar one on vitamin C [108, 153], on which precise information was lacking and immediate action necessary. The League of Nations Technical Commission on Nutrition had, in 1938, estimated the daily requirement of the human adult at 30 mg, whereas in 1943 the United States National Research Council Committee on Food and Nutrition had recommended a daily allowance of 75 mg, an estimate on which the United States Army was basing the food supply to the troops. In wartime Britain it would have been exceedingly difficult to maintain such a high level for the general population, so reliable information was badly needed.

This trial was carried out under the general direction of the Vitamin C sub-Committee of the Medical Research Council (Chairman, Professor R. A. Peters; Secretary, Mr J. R. P. O'Brien). I was again in charge of local arrangements. The experiment lasted from October 1944 to February 1946. One group was given a diet with the lowest possible vitamin C content (somewhere around 1 mg per day) but complete in every other respect. To remove all vitamin C was not possible since it is contained in the majority of foods. Two other groups received the same diet plus a daily supplement of 10–70 mg of vitamin C. The volunteers did not know to which group they belonged, nor did the physicians responsible for the clinical investigations. After 17 to 26 weeks of deprivation all the volunteers on the deficient diet developed signs of scurvy, i.e. skin haemorrhages and gum lesions. The controls remained healthy after 28 weeks. When clear-cut symptoms of scurvy had developed in the deprived volunteers, they were given a supplement of 10 mg vitamin C daily and in all cases this cured the scurvy within a few weeks. The main conclusion reached was that a daily intake of 30 mg allowed a margin of safety and should be adopted as a guide in nutritional policies. This recommendation has since been accepted widely.

My work with the conscientious objectors taught me a good deal about how to get on with people who hold strong views. Conscientious objectors are non-conforming individualists, some indeed are eccentrics who object not only to making war but also to many other things. For the first time in my life I was in charge of a group of people from a great variety of backgrounds and of widely differing attitudes. My research teams had always been small, with the orientation of all its members in the same direction. As a physician, of course, I had been in contact with every kind of person, but the relations between doctor and patient are very different from those which exist between a team leader and the members of a heterogeneous group such as our volunteers.

In my first encounters with them I had not been properly awake to all their sensibilities and I blundered. When trying to explain the importance of exploring the qualities of high extraction flour, I mentioned that if it proved satisfactory for human consumption, its use would save shipping space, since the same tonnage would result in greater nutrition through a fuller use of its cargo. The objectors at once interpreted my words in their own way: saving shipping space would allow the importation of more arms—something which I had not thought of. This *faux pas* of mine, Mellanby wrote, caused a temporary storm. After a short time, however, I began to win the volunteers' confidence and was to keep it throughout our four-year association. It was a surprise to me, on Mellanby's departure, that the volunteers expressed the wish that I should take over the administration of the Institute, although alternatives existed. This made me realize for the first time that I was acceptable as a 'boss' to a largish and not exactly easy group of people. I had never considered myself an easy boss. I knew that I tended to expect a lot from my associates—hard disciplined work and the ability to accept my criticisms. I can only think their confidence in me had something to do with the fact that I have constitutionally a profound respect for every person—whatever his views—as long as he is honest with himself and others. In the course of my association with the team, many problems of human relations came up and I learned how to cope with them. This helped me later when I had to deal with the problems that face the head of a large university department. It taught me in particular the need for close communication between the leader and the members of a team, of explaining one's actions fully, of being accessible, and of giving every member of the team an opportunity to express his views. I learned not to make a major decision without consulting those concerned and yet to take what might be an unpopular decision when I saw this was necessary. I learned to delegate responsibility to those fit and willing to accept it, and to back them if things went wrong. I also learned how to promote a good team spirit—the feeling that the work is not just the leader's one-man show, but everybody's.

The war involved me in several other activities. In 1941 I became a member of a panel set up by the Office of the Lord Privy Seal, Clement Attlee, to enquire into the possibility of increasing the protein supply of the national larder by the manufacture of special yeasts (Torula) which can synthesize protein from low-grade carbohydrate and ammonium salts. This question was considered in several quarters and eventually a 'food yeast' factory was set up in Jamaica where molasses, a cheap surplus carbohydrate, was available. 'Food yeast' was envisaged as a food supplement which would enrich the all-important protein content, but even when added in small amounts it imparted an unpleasant taste. Moreover, its protein content was lower than the nitrogen analyses had suggested; much of the nitrogen was present in the form of nutritionally useless nucleic acids. However, in the 1960s the principle was taken up again, notably by the Lord Rank Research Centre, High Wycombe, England, under the direction of Arnold Spicer. A promising product has been developed. Instead of using yeasts, Spicer experimented with 'micro-fungi', i.e. fungi which have an extensive mycelium but, unlike mushrooms and toadstools, very small fruiting bodies. The mycelium of *Fusarium graminearum* in the presence of ammonium salts readily converts the carbohydrates of grain flour into nutritionally and gastronomically acceptable proteins.

My other war work included participation in nutrition surveys in the Sheffield area. These formed part of a national survey sponsored by the Ministries of Health and Food and were aimed at obtaining information on the state of health of the population as affected by their wartime diet. Another investigation was concerned with the risks of exposure to fluorides in industry, and was carried out on behalf of the Home Office.

In February 1944 I was asked by the Sheffield Teaching Hospitals to take a temporary appointment as supervisor of their Departments of Clinical Biochemistry. This was agreed to by the University as an emergency arrangement and came to an end when qualified personnel became available after the war. I found it a major undertaking because the Departments were in a rather primitive state and needed extensive reorganization.

In my capacity as supervisor of clinical biochemistry I had, in 1946 or 1947, an impressive experience which forcibly brought home to me (and others) the contribution which biochemistry can make to clinical problems. A thoracic surgeon approached me with what seemed to him a puzzling problem; he had removed several oesophageal cancers by oesophagectomy and, though the surgery had been very satisfactory, the patients had died 'without cause' three or four days after the operation. He suspected a biochemical complication.

I discussed the problem with Dr Arthur Jordan, a chemical pathologist who had been appointed after the war, and we came to the conclusion that there might have been a disturbance of mineral metabolism, perhaps a loss of potassium. We decided to investigate the balance of mineral and water metabolism of the next patient to be operated. The result was clear-cut: the patients were not getting sufficient water. After the oesophagectomy they were sedated and, of course, the body's normal regulation of the intake of food and drink was not operating. All the subsequent operations were successful because adequate fluid intake was seen to.

At that time it had not yet been generally recognized that attention has to be paid to water and mineral metabolism in circumstances where normal food intake is no longer a regulatory factor, and especially in patients such as those with oesophageal cancer where the difficulties in eating and drinking may have caused emaciation, dehydration, and loss of minerals before the operation. Nowadays, blood transfusion during the operation, fluid replacement before and after the operation, as well as frequent checks of the level of minerals in the blood (greatly eased by the technique of flame photometry introduced about 1950) are matters of routine. The benefit to the patient from these developments has been enormous.

Despite these other duties during the war years, my research in the laboratory into basic biochemistry never stopped. Towards the end of the war, early in 1945, my group was still very small, with Leonard Eggleston my only semi-permanent (but eventually permanent) technical assistant, and a few temporary collaborators. Although during the war years much of my time was concerned with topical nutritional investigations, the team published some twenty-five papers on basic biochemical problems. These were mainly elaborations of the citric acid cycle and the study of carbon dioxide participation in the metabolism of higher animals, discussed in the next chapter.

11
THE HISTORY OF THE DISCOVERY OF CARBON DIOXIDE FIXATION IN ANIMAL TISSUES

Late in 1939 experiments relating to the study of the citric acid cycle led to a new development: the idea that carbon dioxide, until then thought to be merely a waste product of metabolism, might be an active metabolite taking part in synthetic processes. This turned out to be one of the starting-points of a major development—the discovery of carbon dioxide fixation in animal tissues. Until about 1936 it was thought that carbon dioxide is utilized for syntheses only by a small group of bacteria and by green plants, which are able, in the presence of light, to convert carbon dioxide into organic substances. Indeed, in the last resort, all carbon compounds in plants are derived from carbon dioxide.

That carbon dioxide is fixed and utilized by non-green organisms was first demonstrated by Wood and Werkman (1) in 1936. Studies of the fermentation of glycerol in propionic acid bacteria gave the unexpected observation, from careful quantitative analyses, that the end-product of the anaerobic fermentation (mainly propionic and succinic acids) contained more carbon than had been added in the form of glycerol. Since the medium contained calcium carbonate to neutralize the acids formed during the fermentation the authors suspected that carbonate might have provided the excess carbon. In fact less carbonate was recovered than was originally present and this led Wood and Werkman (2) to the conclusion that propionic acid bacteria are able to utilize carbonate or carbon dioxide and to convert it into an organic compound. In 1938 they showed that the quantities of carbon dioxide used and of succinic acid formed are approximately equimolar. The experimental data conformed with the assumption that glycerol was fermented by the organism by two main reactions, each of a complex nature according to the following schemes:

DISCOVERY OF CARBON DIOXIDE IN ANIMAL TISSUES

(1) $CH_2OH.CHOH.CH_2OH \rightarrow CH_3.CH_2.COOH + H_2O$
    Glycerol   Propionic acid

(2) $CH_2OH.CHOH.CH_2OH + CO_2 \rightarrow COOH.CH_2.CH_2COOH + H_2O$
    Glycerol   Succinic acid

Discussing the significance of the equivalence of carbon dioxide utilization and succinic acid formation, Wood and Werkman considered two possibilities. They suggested

that the formation of succinic acid is by synthesis from a 3-C compound through addition of CO_2; ... pyruvic acid ... may be the point of entry for CO_2.

In 1940 they gave their earlier suggestion a more definite shape by stating

a possible mechanism accounting for CO_2 utilization as well as succinic acid formation involves the addition of CO_2 to pyruvic acid to form oxaloacetic acid, followed by reduction to malic acid, dehydration to fumaric acid and finally reduction to succinic acid (3).

This concept is formulated in the following series of reactions:

$$\underset{\text{Glycerol}}{\begin{matrix}CH_2OH\\|\\CHOH\\|\\CH_2OH\end{matrix}} \xrightarrow{-4H} \underset{\text{Pyruvic acid}}{\begin{matrix}CH_3\\|\\CO\\|\\COOH\end{matrix}} \xrightarrow{+CO_2} \underset{\text{Oxaloacetic acid}}{\begin{matrix}COOH\\|\\CH_2\\|\\CO\\|\\COOH\end{matrix}} \xrightarrow{+2H} \underset{\text{Malic acid}}{\begin{matrix}COOH\\|\\CH_2\\|\\CHOH\\|\\COOH\end{matrix}} \xrightarrow{-H_2O} \underset{\text{Fumaric acid}}{\begin{matrix}COOH\\|\\CH\\||\\CH\\|\\COOH\end{matrix}} \xrightarrow{+2H} \underset{\text{Succinic acid}}{\begin{matrix}COOH\\|\\CH_2\\|\\CH_2\\|\\COOH\end{matrix}}$$

According to this scheme pyruvate is the intermediate which actually fixes carbon dioxide to form oxaloacetate which subsequently undergoes reduction to succinate. In 1940 Wood and his collaborators (4) showed with the help of isotopic carbon dioxide that indeed the utilized carbon is present in the carboxyl group of succinic acid.

The observations which led to the discovery of carbon dioxide fixation were made independently of the work on bacteria at a time when the mechanism of the bacterial reactions was still obscure. But without the knowledge of carbon dioxide fixation in bacteria the idea that this reaction might occur in animal tissues would not have readily suggested itself.

The discovery of carbon dioxide fixation arose from studies of the fate of pyruvic acid in pigeon liver. In 1939 slices of this tissue were found to be capable of synthesizing glutamine in the presence of pyruvate, provided that ammonium chloride had also been added to the pigeon liver preparation (5). As glutamine is also formed by pigeon liver from ammonium oxoglutarate and from glutamate it was thought that these two substances might be the intermediate stages in the synthesis of glutamine from ammonium pyruvate. In 1940 Evans (6) showed that α-oxoglutarate is in fact formed in relatively large quantities when sodium pyruvate is added to suspension of respiring pigeon liver in the absence of ammonium chloride. At this time we already knew, from the work on the citric acid cycle, that animal tissues form oxoglutarate when pyruvate and oxaloacetate, or a precursor of oxaloacetate, are available, the reactions being

$$\text{oxaloacetate} + \text{pyruvate} \rightarrow \text{citrate} + O_2$$
$$\text{citrate} \rightarrow \alpha\text{-oxoglutarate} + CO_2$$

In contrast to other tissues, for example muscle, pigeon liver formed oxoglutarate without the addition of oxaloacetate. Evans therefore came to the conclusion that α-oxoglutarate can arise in pigeon liver in two ways, first according to the citric acid cycle and second by way of an unknown mechanism not requiring the addition of a 4-carbon dicarboxylic acid.

After lengthy deliberations Evans and I decided not to suggest in the discussion of the work a pathway leading from pyruvate to oxoglutarate involving carbon dioxide fixation as Evans was rather sceptical about the possibility of carbon dioxide fixation and because there was no direct supporting evidence. So no positive comment was included for possible mechanisms of α-oxoglutarate formation.

Experiments with isotopic carbon

Evans left Sheffield in the spring of 1940 to return to Chicago, and after his departure we worked independently. It was evident that the question of carbon dioxide participation could be decided only by the use of isotopic carbon but this was difficult to do. Wood and Werkman had used the stable isotope carbon-13 in their work on propionic acid bacteria. Carbon-14 was not yet available and carbon-13 which had been used by Ruben and Kamen (7) had a half-life of only 21 minutes. This meant that experiments had to be completed within one working day after the preparation of the isotope in the cyclotron, the only generator of radioactive isotopes available at that time.

Before asking physicists to help us I decided that we must first obtain supporting evidence for the concept of carbon dioxide fixation. So we demonstrated that the products formed from pyruvate and pigeon liver were exactly the same, and in similar proportions, as those from added oxaloacetate, namely α-oxoglutarate, malate, fumarate, succinate and citrate, and by September 1940 we had come to the conclusion (8) that the only satisfactory explanation for the fate of pyruvate is the assumption that a formation of oxaloacetate from pyruvate and carbon dioxide takes place. Additional evidence was the observation that the rate of pyruvate consumption by pigeon liver preparations depends on the concentration of bicarbonate and carbon dioxide.

These results encouraged me in September 1940 to approach the physicists in charge of the two cyclotrons then in Britain, at Cambridge and at Liverpool. In reply I was told that the apparatus was fully occupied with war work which had to take precedence over everything else. Anxious to follow up the clue I then wrote to Baird Hastings at Harvard University because I had gathered from a publication that his laboratory had the facilities for experimenting with substrates containing carbon-11.

Hastings readily agreed to offer me hospitality in his laboratory, provisionally timed for the summer of 1941. However these plans were overtaken by the publication of Evans and Slotin late in 1940 reporting the successful execution of the experiment which I had planned. It demonstrated conclusively the participation of carbon dioxide in the synthesis of α-oxoglutarate.

It is remarkable that Harland Wood had been working in the very same direction but missed the discovery of carbon dioxide fixation in pigeon liver by a hair's breadth.

An entirely independent approach to the question of carbon dioxide participation in mammalian metabolism was made in 1940 in the laboratory of Baird Hastings, at the suggestion of Birgit Vennesland. When the Harvard cyclotron became available in 1939 Hastings, in consultation with President Conant of Harvard and Professor Kistiakowsky decided to make as much use of the machine as possible by making carbon-11-labelled compounds. They decided to study the fate of labelled lactate and measure the formation of hepatic glycogen and of respiratory carbon dioxide. Birgit Vennesland joined the team as a post-doctoral research fellow in October 1939. At the first formal research conference for which Hastings gathered the team in his room Vennesland, as she put it, diffidently suggested that experiments with labelled carbon dioxide were particularly interesting because there was a good chance that they might be positive. She had arrived at this idea on the basis of her Ph.D. thesis carried out in the laboratory of Martin Hanke in the Department of Biochemistry of the University of Chicago. In this work (10) she made the observation that the obligate anaerobe *Bacteroides vulgatus* depended on the presence of carbon dioxide for its growth (as was already known for a number of other organisms (11)). Though nobody in Hastings's team was optimistic about positive experiments with carbon dioxide in animals it was agreed that such experiments were necessary as a control. The experiments were indeed positive: glycogen was found to contain significant amounts of radioactive carbon introduced into the organism as bicarbonate. The authors pointed out that by assuming a phosphorylation of a dicarboxylic acid formed from carbon dioxide and pyruvate, the conversion of this product into phosphopyruvate and a reversal of the known intermediate stages between glycogen and phosphopyruvate, it would be possible to explain the incorporation of carbon dioxide into glycogen. Subsequent work showed this to be quite a good prediction; though not fully correct, it was broadly so.

Later several other reactions involving carbon dioxide fixation in animal tissues were discovered. A general point which emerges from this account, as from many other stories of discoveries, is that the discovery of the participation of carbon dioxide in the metabolism of heterotrophic organisms was made because the time had become right for it. It was in the path of progress of the elucidation of intermediary metabolism. The tempo of the development was greatly accelerated by the arrival of isotopes. Sooner or later this discovery was bound to be made, and it is no accident that several laboratories were on the same track at the same time.

A fuller account of this subject, which includes the views of the other participants in the development, has been published in *Molecular and Cellular Biochemistry* 5, 79 (1974).

12
THE MEDICAL RESEARCH COUNCIL UNIT
(1944–1967)

Late in 1944 an event took place which was of great importance to the scope of my research. The Medical Research Council, through its Secretary, Sir Edward Mellanby, made an offer to Sheffield University to establish a research unit in my Department, under my directorship. During the war years Government authorities had learned to appreciate the practical values to the nation of science and higher education and towards the end of the war they increased the allocation of funds for universities and for scientific research. The Medical Research Council, whose financial resources depend almost entirely upon the Treasury, decided to use some of their additional money to support promising scientists by providing them with a 'Research Unit'. By this means a chosen research leader could establish a team of collaborators and receive funds for salaries, equipment, and running expenses. The arrangements were long-term and saved the laborious routine of submitting grant applications.

In its negotiations with Sheffield University, the Medical Research Council insisted on the provision of adequate accommodation and insisted also that my status should be raised to that of professor, on the understanding that they would contribute half of my professorial salary. In November 1944, the Vice-Chancellor of Sheffield University, J. I. O. Masson, formulated the outcome of preliminary discussions in the following letter:

My dear Krebs,
 I have replied to Sir Edward Mellanby to the effect that the University would greatly welcome the formation of an M.R.C. Unit of Research in Biological Chemistry with you as Director if his Council decide to propose it. I have added that from the University's point of view it would be essential that our Department of Biochemistry

should continue in full being and should still be headed by you, and that arrangements between the University and the M.R.C. should suitably provide for your dual allegiance.

I have also indicated that we are in process of trying to get you new accommodation and that in the meantime your present quarters, together with the lab occupied by McIlwain, would serve.

The alleged potential danger of 'dual allegiance' was the subject of much argument but I eventually managed to persuade the University authorities that the dangers were non-existent, because the fostering of research which is the task of the Medical Research Council is also a responsibility of a university. This was not the only occasion in my academic career when I had to waste a lot of time on non-problems.

In the initial negotiations (as seen in the Vice-Chancellor's letter) the unit was referred to as 'Unit of Research in Biological Chemistry' but when I saw Sir Edward Mellanby on 12 January 1945, he told me that this designation had only been provisional and was, in fact, undesirable because it could give the erroneous impression that the Medical Research Council was willing to set up Departments of Biological Chemistry or Biochemistry in universities. He asked me to put forward an alternative name which would cover the range of my research activities and yet be somewhat more specific than 'Biological Chemistry' or 'Biochemistry'. I suggested 'Unit for Research in Cell Metabolism' and this was accepted.

At its very start the Unit was a victim of a printing error. When we received the new departmental notepaper carrying the names of both Unit and University Department (my 'dual allegiance') we overlooked for several weeks a misprint. Instead of 'Unit for Research in Cell Metabolism', it read,

MEDICAL RESEARCH COUNCIL
UNFIT FOR RESEARCH IN CELL METABOLISM
TELEPHONE No. 27451
FROM
H. A. KREBS, M.A CAMB., M.D. HAMBURG
PROFESSOR OF BIOCHEMISTRY

DEPARTMENT OF BIOCHEMISTRY.
THE UNIVERSITY,
SHEFFIELD, 10.

It so happened that during the time this paper was in use I had a minor argument with an ex-volunteer of the Sorby Institute. He noticed the misprint on my letter and remarked, in his somewhat quarrelsome reply, 'I note with much satisfaction that you describe yourself as "unfit for research".'

My association with the Medical Research Council as Director of a Unit was a very happy one. I was never short of financial support for the Unit. I was told that when I needed staff, or money for equipment and current expenses, I should ask for it, making out a reasoned case. I could not approach other grant-awarding bodies for support in my capacity as Unit director, though I was free to do so as Head of the Departments of Biochemistry at Sheffield and at Oxford. In those days, applications to the Medical Research Council and drafting them did not take much time. In the case of staff appointments, it was only necessary to submit the credentials of the person to be appointed; in the case of funding, it had to be explained why the money was needed, but no elaborate statements about the research projects were required. So the only limiting factor, both at Sheffield and Oxford, was space. Indeed, only for a very short time, between 1963 and 1967, when the Oxford Biochemistry Department had been substantially enlarged, did I have the ideal amount of space.

A decisive factor in maintaining productivity in research was that within the Unit there was always a nucleus of long-term collaborators, highly skilled, intensely dedicated, and deeply loyal. Some of my early collaborators are still with me at the time of writing in 1980, and my very special thanks go to them. Reg Hems, Derek Williamson, Patricia Lund, and Marion Stubbs have been members of my team for over forty, thirty-three, twenty-three, and fifteen years respectively, with interruptions by military service and work at other centres. Their support over so many years has been invaluable to me. This has been especially true in my 'retirement' work, where their help and understanding have kept me reasonably well on the straight and narrow. Without them I could not have maintained my impetus.

Reg and Derek came to my laboratory straight from school, as technicians. Reg quickly developed into an exceptionally skilful experimenter who mastered to perfection difficult manipulative techniques. He has been my co-author in many papers, contributing to the work not only through his technical skill but also with ideas. Derek, too, soon established himself as an authority in his chosen field and obtained a B.Sc.[17] degree in 1963. He followed this up with the D.Phil. degree in

1967. His standing has been recognized by his election to the Editorial Boards of the *European Journal of Biochemistry* and the *Biochemical Journal*. Pat and Marion joined me as trained technicians, and both studied for their D.Phil. degrees while working in my laboratory.

I have singled out these senior members for special mention because their record of association is exceptionally long and because upon them has rested to a great extent the team's reputation for excellence, helpfulness, good humour, and good teaching of our many students and visitors.

Others who spent many years with the Unit were Robert Davies, Hans Kornberg, Walter Bartley, David Hughes, Ronald Whittam, Rodney Quayle, and Henry McIlwain. John Bacon, although not supported by the Medical Research Council but by Sheffield University as a Senior Lecturer, was also closely integrated with the research team.

All members of the team, through their warmth, helpfulness, friendliness, and expertise, forged links which made the team not only an effective professional body but also a close-knit social unit, indeed a 'family'.

Keeping the team together was made possible by the understanding and help of the Medical Research Council. This was especially important when I moved from Sheffield to Oxford, and from Oxford to my 'retirement' post at the Nuffield Department of Clinical Medicine in the Radcliffe Infirmary. These moves were simple from the administrative point of view once the Medical Research Council had decided to support me, but if I had had to depend on University resources and administration, I would have found it very difficult to keep the team together.

The advantages of having a semi-permanent nucleus in a team are many. Because of the close association and intimate familiarity we got to know each other's strengths and weaknesses, we had no inhibitions about discussing our problems, ideas and difficulties, or about criticizing each other ruthlessly—we knew it was done honestly, in good faith, and in a spirit of helpfulness.

The effectiveness of research much depends on having many different techniques on tap: techniques of chemical and biochemical analysis, of enzymology, and of handling biological material such as mitochondria, perfused organs, and intact animals.

Skills in using, improving, and innovating a variety of techniques and approaches are essential for tackling complex problems. Years of experience in such techniques accumulate extensive knowledge of their pitfalls and of their scope and, of course, lead to improvements. My collaborators not only devised, developed, mastered, and maintained these techniques but also introduced our newcomers (beginners as well as postdoctoral scientists) to what had become routine procedures in the laboratory. Knowing that this was in safe hands left time for myself and other senior members to develop new lines of research and to improve techniques. In a university laboratory, where the training and education of research students is a major responsibility, the teaching of basic techniques is nowadays a very time-consuming task because of the complexity and variety of methods that exist. The team relieved me from a great deal of this work.

Furthermore, in an integrated, co-operative and closely-knit team many new ideas originate and mature as a result of almost continual informal contact in the laboratory. I emphasize 'in the laboratory' because I believe in the value of everyone—say, eight to ten people—working in the same large room. In Warburg's laboratory I had become impressed by the great advantages that can derive from many workers sharing a large laboratory. Many new ideas, solutions to technical snags, even invaluable discussions of everyday problems, have come from being together in one room. I think it is a mistake—a mistake made frequently by laboratory designers—to lock up experimental scientists in small rooms, thus condemning them to long hours of solitary confinement. It inhibits the free interchange of ideas and criticisms which the great majority of scientists desperately need as a stimulus. Some scientists are attracted by the idea of having an area to themselves, an 'empire' where they are absolute rulers. They pay a heavy price. This can be fully appreciated only by those who have worked in a large, shared room. I do not know of a single scientist who, having had the experience of working in the midst of his colleagues, did not become convinced of subtle advantages. True, privacy is needed at times, but this can be found elsewhere, in libraries or at home.

Working together also brings great educational benefits,

especially to the beginner. He is kept in order by his fellow workers and learns to be considerate. He is made to realize that he has to leave instruments in the same clean, tidy, and usable condition in which he finds them. He finds out that it is his responsibility to refill a reagent stock bottle when he has emptied, or near-emptied it. He learns, in general, the discipline necessary in laboratory work.

A large laboratory, apart from providing a stimulating and helpful atmosphere, is also economical in space and equipment, and it increases safety. Expensive instruments such as centrifuges, refrigerators, deep-freezers, optical and counting equipment have to be shared to be fully used; and in chemical work there are hazards from fire, explosion, and flood which are more easily and quickly dealt with when several people are around.

In chemical research there is, in fact, a long tradition of working in large rooms. Liebig's laboratory at Giessen, built around 1840, was the most important single school of organic chemistry of the time and served as a model for modern chemistry departments. It was a very large hall: a drawing by Wilhelm Trautschold in 1842 shows thirteen chemists working in it. Many leading chemists of the nineteenth century trained there and the tradition carried forward. So whenever I had the opportunity to design a research laboratory I planned it on the basis of 'togetherness'. In the new wing of the Oxford Biochemistry Department, completed in 1963, my own laboratory housed some twelve people. Noise and lack of privacy proved to be no problem; at the end of each chemical bench there was a writing desk and this, together with the departmental library, provided all the privacy needed. The continual contact promoted the formation of a socially close-knit team where co-operation, mutual understanding, respect, and affection, with unhampered discussion of every problem under the sun, were the basis of many achievements. And the nucleus was strong enough to transmit its spirit to newcomers.

It has been a continuing source of great pleasure and satisfaction to me that my collaborators have been willing to stay with me over long periods, not only because it has been an enormous help in my work but also for purely personal reasons. I know very well that I have not been an easy taskmaster. I expect high standards of performance and attitude and, at the same time, as some colleagues have occasionally told me, I have not readily given encouragement or praise when members of the team have done well. I took their good work for granted. I have always been fussy when it came to writing up work for publication, often tearing drafts to pieces and insisting on rewriting and yet more rewriting. My own papers have always gone through many drafts in trying to achieve maximum simplicity, clarity, and accuracy.

No one can see himself objectively. Others may be able to explain how, despite my personal shortcomings, I have managed to keep the team together. I have tried to be fair, honest, and helpful, not to demand more of others than I do of myself and above all, I have tried to run the group 'democratically', always discussing problems with them before making decisions. At Sheffield when the team, including the University staff, was relatively small, we had an informal weekly 'business meeting' during a coffee break when everybody from the youngest technician upwards was free to raise a point, ask for explanations, discuss matters of organization, and make suggestions. This was easy at Sheffield because the Unit and the University Department were completely integrated, both starting from scratch at about the same time. At Oxford, the Department had up to 180 members, with many semi-independent sections (each lecturer having his own 'realm') so that complete integration was difficult to achieve. Also, as Head of Department, I had to be very careful to avoid showing favouritism of any kind towards my personal team; but no serious problems arose.

Between 1945 and 1954 (when I moved to Oxford), the Medical Research Council Unit, together with the Department of Biochemistry, published well over a hundred papers in the field of experimental biochemistry.[18] The great majority came from members of the Unit, temporary visitors from overseas and Ph.D. students. The papers covered new aspects of the citric acid cycle, of the metabolism of ketone bodies, glutamate and glutamine, methodological improvements in the microdetermination of several intermediary metabolites, studies of the exchange of potassium between animal tissues and extracellular fluid, oxidative phosphorylation, and the microbiological work carried out by Henry McIlwain and David Hughes. Robert E. Davies carried out outstanding investigations into the mechanism of hydrochloric acid production by the gastric mucosa (1). This research established that when hydrogen ions are secreted in the gastric lumen, equivalent amounts of hydroxyl ions are transferred to the blood circulation. The basic process in hydrochloric acid production is therefore the separation of water into H^+ and OH^-. Much later, in the 1960s, it became known through the work of Peter Mitchell (2) that something analogous to this played a role in the energy transformation in mitochondria.

The move from Sheffield to Oxford in 1954 was a severe test of the solidarity of the team. Family ties and the idea of being uprooted posed major problems for several of the twelve members, but the loyalty to their colleagues prevailed in every single case. Another testing time came thirteen years later when I moved into my retirement job, a job of limited scope and duration (in the first instance it was visualized to last for only three years). In spite of persuasive warnings from the employer, the Medical Research Council, that there was no long-term future in moving with me and that it would be better for them to look for more permanent posts elsewhere, the team again moved as a whole, to my intense gratification. It happened that there were again twelve members though not, of course, the same twelve as in 1954.

At its start in 1945 the Unit consisted of Robert E. Davies (a member of the teaching staff of the Department of Chemistry from 1940–5), Leonard Eggleston (transferred from university staff), and Henry McIlwain and David Hughes, both already on the Medical Research Council staff and attached to my laboratory since 1941. Reginald Hems joined me as a technician on 18 February 1940 and was called up for military service just three years later; he rejoined the team as a member of the Unit when he returned from service in April 1946. People who were members of the Unit at later stages are listed in the Table (p. 140). Included also are those who were not on the Medical Research Council payroll but who, in every other way, were members of the team.

Young graduates joined the Unit in the first instance to work for the Ph.D. degree; the best among these could be retained as staff members after they had completed their degree work. In addition there were visitors who spent fellowships or sabbaticals with the Unit. The only difference between the University

Long-term members of the M.R.C. team

The Table includes those with a minimum association of five years. It therefore excludes students reading for higher degrees and visitors who came for shorter periods. The associations with Hems, Burton, Whittam, Williamson, Lund, and Stubbs were interrupted by national service or by periods spent at other laboratories, though all returned to the fold.

Name	Period of Association	Present position
L. V. Eggleston	1936–1974	Deceased
R. Hems	1940–present	
H. McIlwain	1941–1947	Professor of Biochemistry, Institute of Psychiatry, University of London
R. E. Davies	1945–1956	Professor of Biochemistry, University of Pennsylvania
D. E. Hughes	1945–1964	Professor of Microbiology, University of Wales at Cardiff
H. L. Kornberg	1945–1961	Professor of Biochemistry, University of Cambridge
D. H. Williamson	1946–present	
W. Bartley	1946–1963	Professor of Biochemistry, University of Sheffield
J. S. D. Bacon	1948–1954	Head of Department of Carbohydrate Chemistry, Rowatt Research Institute, Buckburn, Aberdeen
A. Renshaw	1949–1967	Senior Technical Officer, Medical Research Council Clinical Research Centre
K. Burton	1949–1967	Professor of Biochemistry, University of Newcastle
R. Whittam	1951–1963	Professor of Physiology, University of Leicester
A. Gascoyne	1954–1961	Technician, Department of Biochemistry, University of Oxford
B. M. Notton	1954–1966	Administrative Officer, Department of Biochemistry, University of Oxford
J. R. Quayle	1956–1963	Professor of Microbiology, University of Sheffield
P. Lund	1957–present	
E. A. Newsholme	1962–1967	Lecturer in Biochemistry, University of Oxford
P. C. Gregory	1964–1977	Left science
M. Stubbs	1965–present	
C. R. Farrell	1965–present	
V. Ilic	1967–present	
D. Wiggins	1971–present	

staff and Unit members was the distribution of teaching responsibilities. Members of the Unit did participate in undergraduate teaching, but on a smaller scale than the University staff. All took a share in the administrative responsibilities.

A major threat to the work of the Unit developed in 1959 when the University Grants Committee in their Report on University Development suggested that the responsibility for some Medical Research Council units should be transferred to universities. This policy was expressed by the statement: 'A line of original scientific work which has proved itself should be financed in the ordinary way from the general University income and the University should accept full responsibility for it instead of regarding it as an extraneous activity.' In compliance with this policy, the Medical Research Council asked Oxford University to enter negotiations about the transfer of the team. The letter, from R. H. C. Cohen, Deputy Chief Medical Officer of the Medical Research Council to Sir Folliott Sandford, the Oxford University Registrar, stated

This Unit could now be considered as having become an integral part of its parent Department. The Council indeed felt that of the Council's seventy or so units and groups, this Unit in particular provided one of the more clear-cut examples of Research-Council-aided projects that could be regarded as 'established and ripe for transfer'— in the words of the University Grants Committee's report.

At the same time, the Secretary of the Medical Research Council, Sir Harold Himsworth, assured me that his Council was not in any way trying to get rid of my Unit. On 14 January 1959 he wrote to explain the attitude of the Council:

You may have noticed in the report of the University Grants Committee that the question was raised of universities taking over Medical Research Council units that had become fully established and an integral part of the life of a university. Of course, such a suggestion could apply only to a few of our units: those that are engaged on fundamental work which integrates naturally with university activities. Recently the Council carried out a review of all their units to see which, if any, might be considered from this point of view. The result was that they picked out five and instructed me to look into the matter; and one of the five they suggested was yours.

I would hate you to think that the Council were in any way trying to get rid of your unit. On the contrary, they were rather chagrined to find that the five units that had grown most to be part of the university where they were situated were all units of which they were particularly proud, and which they would be reluctant to part with. But the Council had been a party, in principle, to the views that the University Grants Committee put out in their report. They recognised that some such development as that foreshadowed was in the natural history of the evolution of a unit, and that in the long run it was probably in the best interests of the subject to make arrangements which gave it continued security in the university beyond the tenure of a single director.

In the subsequent discussions with the General Board (the University Authority which controls academic matters) it became clear that transfer of the Unit to the University would meet with major difficulties. The General Board insisted that people to be transferred should be paid according to the University salary scales for lecturers and research officers, which were—and still are—tied strictly to age and not related to merit.

In principle, the scales used by the University and the Medical Research Council for full-time staff were very similar. The Government, which through the Treasury provided the funds both for universities and Research Councils, intended that similar work should be similarly rewarded, but their rulings referred to broad ranges, not to details such as the age-tie.

When the transfer of the Unit was being discussed, it became clear that serious salary problems would arise over only three of its senior scientists out of a total scientific, technical, and clerical staff of twenty; but these three were the key members: Burton, Kornberg, and Quayle. All three had outstanding qualities which were later recognized by appointments to Headships and election to the Fellowship of the Royal Society. Without such people the quality of the work of the Unit would not be maintained.

There were other important aspects militating against a transfer. To replace the Medical Research Council support, the University Grants Committee intended to make an earmarked extra allowance to the University, but this would be for only one five-year period. Thereafter the extra allocation would be absorbed into the general University budget. Moreover the earmarked allocation would not have the flexibility of the

Medical Research Council's support, which allowed me at any time to put to them a proposal for consideration. As I shall explain in a later chapter, Oxford was for me a splendid place, but only because I was, in respect to research, entirely independent of the University for financial support. Not only was my own research and that of the Unit financed from outside sources (the Medical Research Council, the Rockefeller Foundation, and the United States Public Health Service were very generous) but I also managed to channel large sums of research money into the Department as a whole.

That scientific research should be largely financed from sources outside the University is reasonable and has become a widespread practice in the Western world during the last twenty or thirty years. A university may lack the expertise for assessing the merits of research proposals and, in a democratically governed university where action is taken on decisions made by a majority of senior members, it can often be embarrassing for colleagues to have to decide who, amongst themselves, merits support and who does not. Remote control by a body of proved experts, real peers, is far better—as long as the control is essentially in the hands of practising fellow scientists of high standing and not of administrators who cannot form first-hand opinions of the work.

So I have no quarrel with the fact that I had to collect my own research funds. But that the University should lay down rules about salaries, penalizing those who dedicate their time and energy mainly to research is very wrong. A university should aim at equally high standards in teaching and research. Much university research is, after all, advanced teaching at the postgraduate level. Most members of my team probably spent more time on this kind of teaching than did any college Fellow on undergraduate teaching.

Another argument raised against incorporation of the Unit into the Department of Biochemistry was the claim that it would disturb the balance of activities within the Department; it would put too much emphasis on a narrow field. That this was a feeble point is forcibly illustrated by the fact that members of the team later became professors not only of biochemistry but also of physiology, microbiology, zoology, and clinical biochemistry, and that post-doctoral collaborators included nephrologists, cardiologists, and an anaesthetist of professional standing.

In the end I felt that the University as a body—as opposed to many individual members—attached little importance to my Unit's work, which incidentally was not only to advance knowledge but also very much to bring forth the next generation of academic leaders.

So no compromise seemed possible, and after two and a half years of argument and counter-argument and an enormous amount of wasted paper and committee work, the negotiations were abandoned. On 6 July 1961 I wrote to Dr Cohen,

> I believe that the basic obstacle is a major difference in the nature of Medical Research Council appointments and Oxford University appointments. The Council looks upon its staff as full-time workers. They are expected to work a minimum of 39 hours, to take no more than six weeks' leave per annum and they must not use their so-called leisure time for paid appointments except by special permission. For the service expected, the Council intend to pay 'living' salaries commensurate with those earned in other professions. The University, on the other hand, although it specifically designates Demonstrators by statute as 'full-time' staff, does not effectively formulate any standards of service. In practice it has almost none.

Dr Cohen replied on 28 July,

> We reported the whole position to the Council at their meeting last Friday and they agreed that the continuation of the Unit in its present form until your retirement from the Chair was the only possible course. They were sorry that the situation at Oxford had prevented the implementation of what you yourself, I believe, once described as a 'basic principle of government policy'; but at the same time they were very happy to know that they were going to be associated for a further period with the first-rate research which was being undertaken under your direction.

The failure to reach a satisfactory outcome stemmed from the fundamentally different outlook of Oxford's General Board and the Research Councils. The Councils expected first-rate research and were willing to offer first-rate salaries for excellence. They realized that first-rate scientists were not to be had at second-rate salaries. The General Board, though paying lip-service to research, refused to recognize excellence in research financially. As a consequence many a scientist left Oxford and the country.

A foreseeable problem to be faced was the final fate of members of the Unit when I had to retire from the Chairmanship of the Department of Biochemistry. I always felt confident, however, that this would resolve itself easily, provided that they were first-rate scientists. In the event, as the time of my retirement approached, all the senior members were well placed in senior academic positions elsewhere. A small group of the more junior people moved with me into my retirement work, which was made possible by the generosity and understanding of the Medical Research Council.

By the time the Unit had to be disbanded, all the senior members—those who had stayed for prolonged periods—had obtained professorships or equivalent posts:

Robert E. Davies	University of Pennsylvania
Hans Kornberg	first at Leicester, later at Cambridge
Walter Bartley	Sheffield
Rodney Quayle	Sheffield
Kenneth Burton	Newcastle
Ronald Whittam	Leicester
David Hughes	University College, Cardiff

At the time of writing, no less than six out of a total of about a dozen who were senior members of the Unit have been elected to Fellowship of the Royal Society—Davies, Kornberg, Burton, Whittam, Quayle, and Gordon Dixon. During the same period only two members of the Oxford Biochemistry Department, Alexander Ogston and Keith Dalziel, were elected to the Fellowship, out of a complement of about fifteen. I do not include Donald Woods and Joel Mandelstam who had done most of the work which qualified them for the Fellowship before they came to Oxford, i.e. while they held full-time research appointments under the Medical Research Council.

The small number of Lecturers in the Department who were elected to the Fellowship does not mean that they were less gifted or worked less hard than the members of the Unit; I believe that a major reason for their lack of distinction in research was the fact that they did not have sufficient time to devote to it. This in turn was not a matter of unduly heavy departmental teaching duties, but was due to the fact that they held very time-consuming but well-remunerated second appointments as tutorial College Fellows. Keith Dalziel recognized the importance to major research of having plenty of time for it and steadfastly refused to be considered for tutorial Fellowship—at considerable financial sacrifice. It was a source of great satisfaction to me that this sacrifice was rewarded by his election to the Royal Society. Ogston, although very busy as a Tutorial Fellow, succeeded in making a major discovery of a theoretical kind, which I discuss in Chapter 14. It was not excessively time-consuming. According to Ogston himself, 'It was a matter of seconds'.

Until the Second World War, many tutorial Fellows managed to combine first-rate research with College duties, but since that time research has become much more exacting in terms of time because of the increasing complexity of the subject-matter.

The separation of research and teaching is, of course, highly undesirable but it seems to me essential that teaching should not be better rewarded financially than research.

1. The author's parents at the time of their engagement in 1894.

2. Hildesheim, the author's birthplace, as it was before it was bombed in 1945.

(a)

(b)

3. (a) The author as a student, about 1922; (b) (*right*) at the congress of Internal Medicine in 1932, where he made his first report on the ornithine cycle of urea synthesis.

4. F. G. Hopkins (Cambridge) and T. Thunberg (Lund, Sweden) at an international conference in 1936.

5. Nobel Prize-giving ceremony. Lady Churchill speaking to the Krebs family.

6. The author lecturing at Lindau in June 1966.

7. The author with his team at the time of his 'retirement' in 1974. *Key*: 1. Rose Farrell; 2. Bob Harris; 3. Marion van den Berg; 4. the author; 5. Morag Stuart; 6. Mary Thompson; 7. Philip Gregory; 8. Madge Barber; 9. Marion Spry; 10. Vera Ilic; 11. Patrick Vinay; 12. Jane Holloway; 13. Brendan Buckley; 14. Barry Tyler; 15. Pat Lund; 16. Derek Williamson; 17. Reg Hems; 18. Neal Cornell; 19. Ed Prosen.

8. The author (*centre*) with (*left to right*) Ronald Estabrook, Severo Ochoa, Konrad Bloch, and Carl Cori at a conference in 1980.

13
RE-ESTABLISHING CONTACTS WITH GERMANY
(1945–1949)

In 1945 the husband of my secretary Dorothy Austen was serving as an army chaplain with the British Forces in Germany. Early in April he was with the advancing army at Celle, only forty miles from Hildesheim. Through his wife I sent word to him that I would be most grateful if he could try to make contact with my stepmother and half-sister, Maria and Gisela, at the old address, Zingel 9. The message reached Canon Austen just before the German surrender on 7 May. The next day, 'VE Day', he was officiating at a thanksgiving service on Lüneburg Heath but on the following day he was free and set off by car for Hildesheim, about seventy miles. The journey was not easy because roads and bridges had been destroyed. We heard from him on 16 May.

Two blown bridges were one each side of a big town [Hanover] which we have often bombed, about 20 miles from 'journey's end'. We did not go in through the town, but what we saw of it was badly knocked about, though not as badly as 'journey's end'. There was no one living in the centre of 'journey's end' I should think, and I did not see a whole house. I did not attempt to find Zingel 9 but went in search of the Burgomaster. I duly found his office, but the two clerks knew nothing of the Professor's stepmother until one of them remembered that after Zingel 9 had been bombed, she had shared a house with a Dr. Osterwald. So I was taken along to see him and after about ten minutes of talk in which he said that he had no idea where she was living—or even if she was living—his son arrived and was able to tell me where I should find her. So off I went again, about five miles to the village of Machtsum, and there I found her and her daughter, Gisela, living in comfort in a lovely farmhouse.

She nearly fell on my neck when she knew why I had come and indeed she guessed the purpose of my visit before I stated it. She had had a thin time, as the Nazis would not allow her to teach and would not allow her to bring away any furniture from her house when it was bombed a second time. But she had some good friends who had helped her through. I stayed about an hour and suggested that she should write the enclosed note, which you will doubtless pass on to the Professor. He cannot answer yet, so I will not send his stepmother's address yet. I have it, and will send it on as soon as correspondence is allowed, which I imagine will be before long. I doubt if I shall have the opportunity to go and see her again as I am now so far away, so he must possess his soul in patience till he can write. If it is possible through the Red Cross, would you let me know and I will send on the address. You can assure him that both mother and daughter are very well, and that the mother will probably be teaching again soon, through the kind offices of the Bishop of Hildesheim.

The farmer who had offered shelter after the complete destruction of the Krebs home had been a grateful patient of my father.

One can imagine the dramatic situation at the farm when a British officer arrived out of the blue only two days after the end of hostilities to enquire about my relations, able to tell them about myself and my family and able to pass a message to us in Sheffield.

So, after a separation of over four years, family contact was re-established. Until the very last Maria and Gisela had been persecuted by the Nazi authorities, proscribed and threatened with deportation to concentration camp. Luckily, one of the Nazi officials had also been an appreciative patient of my father and showed them a little humanity. When he was later arraigned as a war-criminal, he pleaded in his defence that he had shielded my relatives.

At the farm it was clear to Maria and Gisela that they were being sheltered against the will of the farmer's wife, who was afraid of Nazi reprisals. My stepmother could not help but hear voices raised in loud and strident argument about their being there. People in the village were also uneasy, but kept silent. But from the moment of Canon Austen's visit attitudes changed. They were begged to stay in the village and to protect the farms from military requisitioning to accommodate Allied soldiers.

It was a long time before normal postal services with Germany were resumed. Understandably the Allies were in no mood to foster friendly relations. Fraternization between army personnel and German civilians was forbidden. My contacts with my relatives depended upon the good offices of members of the Armed Forces and of the Military Government; fortunately, Hildesheim was in the British zone of occupation and we managed to exchange letters occasionally.

It also took a very long time for normal contacts with German scientists to be resumed. The first to contact me was Theodor Bücher, like myself a pupil of Otto Warburg, first in Berlin and later (after the evacuation of the laboratory) at Liebenburg, some thirty miles to the north. In 1944 Bücher had completed four important pieces of work, including two on enzymes of glycolysis. They had been accepted for publication by the *Biochemische Zeitschrift* and had reached the proof stage, but they were lost when the printing establishment in Leipzig was destroyed by bombs.

As the Russian army advanced on Liebenburg, Bücher moved his family westward to the British zone of occupation and settled in a village not far from Lübeck. He got work in a chemical factory. One day, members of the British Intelligence Service happened to see him dismantling some electronic equipment and suspected him of trying to set up radio communications. Bücher was taken for interrogation but succeeded in convincing his questioners that he was not a secret agent but a research scientist who, moreover, was anxious to have their help in getting three of his unpublished papers to Britain. He asked if they could be sent to Professor Donnan, care of the Royal Society. Through Donnan, these papers reached me at Sheffield and once again Canon Austen was instrumental in keeping Bücher and myself in touch. I submitted the papers for publication to the *Biochemical Journal* and received a letter, dated 22 January 1946, from the Chairman of the Editorial Board, stating, 'The Committee of the Biochemical Society have directed the Editorial Board that, until further notice, the Editors should not accept papers submitted from ex-enemy countries. In the circumstances it will obviously be better for Dr Bücher to publish in a neutral country or to wait until publication begins again in Germany.'

In the event they were published not in a neutral country, but in Holland. This came about because I had an opportunity to submit the papers to the new journal, *Biochimica et Biophysica*

Acta, and in July 1946 they were accepted for publication through the courtesy of the Editor, Professor Westenbrink and of the publishers, the Elsevier Company. The papers appeared in the first volume of *Biochimica et Biophysica Acta* and were widely recognized as being of great importance.

The next German colleague to get in touch with me was Kurt Henseleit, my collaborator in the work on the ornithine cycle. He wrote some twenty-one months after the end of hostilities, a delay partially due to the fact that he had been a prisoner of war in the American zone. He was anxious to make contact with me but did not know my whereabouts. In February I received the following letter, written in faltering English and addressed to the Royal Society of Medicine.

To: The Royal Society of Medicine, London.
I am searching Mr. Dr. med. Hans Adolf Krebs who emigrated from Germany to England in 1933. I mean he has gone to the Institute of Mr. Prof. Hopkins.
I should be very much obliged to you for getting his address. I have made with him 'Die Harnstoffbildung im Tierkörper' in the Clinic of Prof. Thannhauser at Freiburg i. Br. in 1932.

I was very pleased to hear from Henseleit and answered his letter immediately; we remained in contact until his death in 1976.

My first news of Otto Warburg came from the letters from Theodor Bücher, from whom I learned that Warburg had come through the war unscathed though he had lost all his research equipment. It had been removed by the Russian Army and his efforts to recover it were fruitless.

I have no record of any correspondence with Warburg from 1939 until 1947, when he sent me a copy of a book he had written during a period of compulsory absence from the bench, entitled *Schwermetalle als Wirkungsgruppen von Fermenten* (1). It was a brilliant summary of his achievements and views on the role of heavy metals in cell metabolism, but it was peppered with fierce and, I thought, unfair polemics, especially against my close friend David Keilin and others such as Heinrich Wieland and Richard Willstätter. I felt strongly that I had to take issue with his style of controversy, especially in defence of David Keilin. So on 14 June 1947 I wrote to him,

Lieber Herr Warburg,

I wish to thank you for sending me a copy of your book which I have studied with the greatest interest.

I hope, however, you will forgive me if I offer some critical comments on certain aspects of the book. I find the pleasure of reading the account of your magnificent contributions to science marred by some very ghastly polemics. In many instances your remarks are unfair because you misrepresent the views of your opponents, and they are discourteous because you portray your victims not only as stupid and muddle-headed, but also as dishonest people who try deliberately to confuse the issue. I think it is a great pity that you were not kinder and more generous towards your fellow scientists and that you tried to cast ridicule and contempt upon them.

In the old days you kept polemics deliberately out of your scientific papers and put them into special notices. Moreover, their wording often betrayed an amusing sense of humour and people fancied your attacks were not meant quite seriously. Now you intersperse your scientific account with personal and unscientific remarks and your attacks sound bitter; one can hardly regard them as good-humoured jokes. There is really no reason for feelings of bitterness towards your fellow scientists who, all over the world, acknowledge you as one of the greatest contemporary scientists.

Is it the agonising environment of the last 14 years, and your loneliness, that has affected your attitude? Or has *my* attitude changed and have I become over-sensitive towards your sort of polemics? In the world in which I now live, as you will know from the modern Anglo-American or French literature, polemics are very rare and they are never discourteous. Scientists here are content to rely on the rule that scientific untruths will sooner or later die a natural death. They find it quite unnecessary to assist too eagerly the maturing of the judgement of time.

May I express the wish that you omit from your second book all those remarks that are tinged with unscientific emotions, so that your real work will stand out unblemished in all its splendour? To put this wish to you is, in fact, the prime object of writing this letter.

Mit bestem Gruss, Ihr

I received no answer. Some time later, in 1949, Dean Burk, an External Member of Warburg's laboratory who was in regular contact with him, passed on to me a comment Warburg had made: 'Krebs', he said, 'did not like my book and wrote to criticise my way of correcting the errors of Keilin and Wieland. He wrote, "We Anglo-Saxons do not do that sort of thing"!'

So I sent a copy of my above-quoted letter to Burk and he promised to correct the distortion. In return I had a postcard with the message 'All forgiven if not forgotten', and signed jointly by Burk and Warburg as 'We poor Germans'.

News of other friends, colleagues, and acquaintances slowly trickled through, and I received letters from Butenandt, H. H. Weber, K. Gollwitzer-Meier, Tidow, and others. I learned from Professor Parnas of Lwow, then living in Moscow, that Pawel Ostern by going into hiding had survived the murder of ninety-eight professors in Lwow on the first day of German rule, but that two years later he had been denounced. When he was about to be arrested he had taken cyanide rather than die at the hands of the Nazis. Many of the letters painted a sad picture. Laboratories had been bombed; there were no books, no supplies, no money, and little sympathy from the occupying forces. Until 1947 there was hardly any scientific literature. Drugs were in short supply and diabetics died for lack of insulin.

While personal contacts with German colleagues gradually developed after 1946 when postal services were re-established, official contacts at the level of the scientific societies were almost non-existent until 1949. This question became acute in connection with the organization of the First International Biochemical Congress to be held in August 1949 in Cambridge. I was a member of the Biochemical Society Committee at this time and had learned by chance that the organizers of the International Physiological Congress to be held in 1947 in Oxford had, for the first time, decided not to include biochemists among those to be invited. I notified the Biochemical Society Committee about this change in policy and they appointed a subcommittee charged with the preparation for the setting-up of an international biochemical congress. The subcommittee approached sister societies abroad to find out whether they would welcome such a move, and if so, whether biochemists from ex-enemy countries should be invited to attend. In 1948 the subcommittee came to the conclusion that because of the objections they had received from continental biochemists—understandable at the time—they should bar Germans domiciled in Germany.

I felt very strongly that this kind of exclusion should not be allowed to continue—three years after Germany's surrender. I believed it not only morally wrong to cold-shoulder decent Germans, but also a great political mistake because it would fuel pro-Nazi elements in Germany. I also felt that there was a special reason why I in particular should take a stand against this discrimination. If I did not, my fellow committee members might assume that as a refugee I might have strong feelings against any resumption of relations with Germany. In a sense I felt that I was in a special position to advocate the re-establishment of normal relations with ex-enemy countries. Frank Happold strongly supported my plea and the Committee as a whole was sympathetic. Further negotiations with continental societies led to a compromise and it was agreed to invite a small number of selected German biochemists after careful screening of their political histories. The outcome was the sending of a personal invitation to four German biochemists: Kurt Felix, Emil Lehnartz, Theodor Wieland, and Hans-Joachim Deuticke. Benno Hess managed to attend informally as friends had invited him to come to England at that time. The effects of the Germans' participation were quite out of proportion to their small number, because once the principle had been established, contact with German biochemists rapidly became normal.[19]

The complete severance of contact between scientists of warring countries was a consequence of the twentieth-century concept of 'total war'. Until then, war did not necessarily interrupt contact between intellectuals. Even in the middle of the Napoleonic wars Sir Humphry Davy and his wife visited France, travelling in their own carriage, to meet French scientists in Paris, among them Ampère, Berthollet, Cuvier, Chevreuil, Gay-Lussac, and Laplace. In contrast during the Second World War no personal contact was allowed. So deep was the sense of hostility in Germany that the family of the famous medical scientist, Ludwig Aschoff, was amazed by the objective, fair, and warmhearted obituary of Aschoff which appeared in the *Lancet* (2) in 1943, written by Dr Alistair Robb-Smith of Oxford. This had reached Germany via Switzerland and caused surprise because it made it obvious that no personal enmity existed.

I should perhaps explain further the frame of mind which prompted me to support the resumption of normal relations with Germany. I have often been asked what I as a refugee felt about the country of my birth, where incredible crimes against humanity, and against my closest relatives and friends, had been committed by a government which had the support of large numbers of people. My sentiments were in no way dominated by a nostalgia for my original home. Much as I had loved Hildesheim, my early liking for England and its people had quickly deepened to the point of feeling it to be my real home. But it never occurred to me to identify all Germans with Nazism and hence to accuse or cold-shoulder every individual. My personal faith in the decency of the majority of them was greatly helped by a number of personal experiences, some of which I have related in this book. I knew that most of them were no less decent—and no more heroic—than people elsewhere. I knew that circumstances had forced them to yield helplessly in the face of immense pressures and intimidation from Nazi power. If they were to survive, they had to identify themselves for all practical purposes with the regime, and serve it. The Nazis used fearful threats of reprisals against the families of individuals who showed signs of opposition. Thus Adolf Butenandt told me that when he was awarded a Nobel Prize in 1938 he was asked to sign an abominably rude letter (written by party officials) to the Nobel Foundation, refusing the Prize. When he indicated an unwillingness to sign, he was reminded by the Gestapo that not only he but his wife and five children would suffer. Tens of thousands of Germans paid with their lives for their attempts to oppose the Nazi regime, but the majority—not being heroes—felt entirely unable to stand up for their true beliefs and feelings.

The results of the terror of the Gestapo gave distant observers the impression that the whole of Germany supported Hitler. When he came to power he certainly had a huge following, although he did not achieve his absolute power through a parliamentary majority. After the war it was difficult for the Allies to discover people who would admit to having been pro-Nazi, and it will never be known what percentage voluntarily supported Hitler. My guess is that at some stages it may have approached or exceeded half of the population. But to be anti-German seems to me just as bad as being anti-Semitic. For many refugees it was impossible to look at things this way; the cruelties, killings, robberies, and atrocities which they had witnessed made it impossible for them ever again to set foot in the country of their birth.

My own first visit came at the end of August 1949, after an absence of more than sixteen years. I had received repeated warm invitations from German biochemists to visit them, and was on my way to participate in the annual meeting of German physiologists and biochemists at Göttingen.

Travelling by train via the Hook of Holland, I found that entering Germany moved me more than I had anticipated. As the train sped along, I found myself unable to read or study as I usually do on train journeys and instead I gazed at the scenery, many features of which had been so familiar to me, had been a part of me as my home background. At the same time, the knowledge of the history of the past sixteen years affected my feelings, especially towards the people. Who, among these ordinary-looking people, had welcomed Hitler and followed him obediently to the end? Who among them had been a member of the hateful Gestapo? Who among them had perpetrated the misdeeds which Hitler had directed? Who among them had taken an active part in the horrors of the extermination camps where more than twenty of my close relatives had been put to death?

The mixture of emotion at homecoming and sense of estrangement, even revulsion, evoked by the Nazi period caused in me a disturbance which had an eerie quality and which faded only gradually during my seven-day stay.

There was much devastation in the large towns, particularly in Osnabrück and Hanover, but this I did not find particularly upsetting; it was expected and only quantitatively different from what was to be seen in England. Göttingen had not suffered serious bomb damage and little seemed to have changed since my student days in 1918 and 1919.

At the Congress I was received in a most friendly, at times almost enthusiastic, manner by colleagues, several of whom I had known well and who seemed to take a genuine pleasure and satisfaction in seeing me again. Some spoke freely and with feeling about the atrocities and stupidities of the Nazi regime and all emphasized how much they were suffering from their present isolation and how anxious they were to establish foreign contacts. At first I stayed deliberately rather reserved, listening rather than talking. Then I was asked to chair a session, and because some of my old acquaintances seemed to me to be glossing over the events of the Hitler years as if nothing unusual had happened, I took the opportunity of being in the Chair to remind the audience of what had happened to me and many others like me. At the end of the meeting the Congress organizer referred to my remarks, expressing appreciation that I had attended in spite of what had occurred in the 1930s. This evoked thunderous applause.

I found time to look at Göttingen, which I had not seen since my students days. Although many features of the town and people were familiar, I felt like an alien in the country where I had spent the first thirty-two years of my life.

When the Congress was over I went to Hildesheim to visit my stepmother, Maria. I had heard so many descriptions of the town's complete destruction that I was not shocked by the ruins. On the contrary, I had not expected to find almost all the streets cleared and passable. They were crowded and busy with people and rebuilding was proceeding at great speed. The centre of the city was still a desert of ruins, only the shells of some of the larger buildings still stood, but work had already begun on rebuilding the Rathaus and some of the churches.

The house in which I had grown up was a heap of rubble, much of it already overgrown by trees, bushes and weeds. A big lime tree stood almost unchanged, just slightly askew. A gingko tree had apparently benefited from the increased light and air after the destruction of the house; it had grown enormously and was much taller than the three-storey house had been. Most of the garden was covered by self-sown bushes, although some fruit trees were still flourishing. I was amazed to see that much of the old vegetation had survived the fire and blast whilst the iron of the gate and fences was but a twisted mass. There was nothing to remind me of the inside of the house; everything had been destroyed.

By now, Maria held a busy key position in the education department of the area government offices, dealing in particular with the problems of 'de-Nazifying' the education system.

I spent some time walking in the hills and woods, enjoying their familiar and unchanged beauty and experiencing a strong sense of attachment to the haunts of my youth. I called upon one of my schoolteachers, Johannes Gebauer, for whom I had a special feeling of attachment because of his effectiveness as a teacher and because of his uncompromising and forthright integrity. He was a distinguished historian who had written a standard book of the history of Hildesheim, and had resisted Nazi policies. Some eighteen years later, when I received the honour of the Freedom of the City of Hildesheim, I learned that Gebauer was the only other Freeman at that time.

The charms of the medieval town I had known before the war had completely disappeared. I also found a big change in its people. Even before the war, the Nazis were moving large numbers of people from one part of Germany to another to man huge munitions factories. Some of these had been built in the extensive woods near Hildesheim; the largest was a branch of the Bosch Works, which had involved the transfer of many workers and their families from the Stuttgart area. Then the destruction of the town had forced many inhabitants to make their homes elsewhere. After the war, refugees had flooded in from the east. Thus the population was very different from the indigenous one and few spoke in the old, down-to-earth way or used the characteristic turns of phrase I knew. The language of the majority was colourless, often rather ugly and slovenly. It was yet another change to intensify the feeling that I no more belonged in the place of my birth.

14
FINAL FORMULATION OF THE CITRIC ACID CYCLE

At the end of Chapter 9 I mentioned that experiments with radioactive carbon in 1941 led to the belief that citric acid was not an intermediate in the cycle, but that in 1948 Alexander Ogston discovered, by a penetrating theoretical analysis, that this belief was wrong. This discovery by Ogston (whom I had first met in the mid-1930s soon after my arrival in England and who became a close colleague and friend in the Department of Biochemistry at Oxford), I will now describe in some detail for scientific readers. It turned out to be a matter of great importance to the progress of biochemistry. The story brings home that in biochemistry—as has long been appreciated in physics—theoretical studies are no less important than new experimental data.

The erroneous belief that citric acid is not an intermediate arose from the interpretation of the fact that citric acid has two identical $-CH_2.COOH$ groups attached to the 'central' carbon atom:

$$COOH.CH_2-\underset{COOH}{\overset{OH}{C}}-CH_2.COOH \qquad \text{Citric acid}$$

It was taken for granted that the two identical groups have an equal chance of forming isocitric acid in the presence of the enzyme aconitase and that two forms of isocitric acid would therefore arise:

FINAL FORMULATION OF CITRIC ACID CYCLE

$$\text{COOH.C(OH)(H)—C(H)(COOH)—CH}_2\text{.COOH}$$

$$\text{COOH.CH}_2\text{—C(H)(COOH)—C(OH)(H).COOH}$$

Two forms of isocitric acid

The two forms differ in that the OH group of citric acid has moved to the left in the upper formula and to the right in the lower one. This was expected because it was thought to be a general law of chemistry that identical groups of a molecule behave identically in chemical reactions. Before Ogston's discovery it was not appreciated that the 'law' does not necessarily apply to reactions which are catalysed by enzymes.

It therefore came as a great surprise when experiments with isotopic carbon (carbon-14) indicated that the two –CH_2.COOH groups did not react identically. Experiments proving this were carried out in 1941 by Wood, Werkman, Hemingway, and Nier, and by Evans and Slotin, and concerned the synthesis of α-oxoglutarate from pyruvate in pigeon liver. This synthesis involves as a first step the formation of oxaloacetate from pyruvate and CO_2, as already briefly referred to in Chapter 11. The formation of oxaloacetate is followed by the reactions of the citric acid cycle. The relevant intermediate stages are indicated by the following formulae where the isotopic carbon is marked with an asterisk. The oxaloacetic acid formed from pyruvic acid and carbon-14 carbon dioxide reacts with acetyl-CoA to form citric acid:

$$\star CO_2 + CH_3.\underset{CO_2H}{\overset{O}{\underset{\|}{C}}} \qquad \text{Pyruvic acid}$$

$$\downarrow$$

$$\star CO_2H.CH_2.\underset{CO_2H}{\overset{O}{\underset{\|}{C}}} \qquad \text{Oxaloacetic acid}$$

$$\downarrow \quad CH_3.CO\text{–Coenzyme A} \qquad \text{Acetyl–CoA}$$

$$\star CO_2H.CH_2\underset{CO_2H}{\overset{OH}{-\underset{|}{C}-}}CH_2.CO_2H \qquad \text{Citric acid} + \text{CoA}$$

The two –CH_2.COOH groups of citric acid—which are indistinguishable as far as their ordinary chemical reactivity is concerned—have now become distinguishable by the difference in radioactivity, the .CH_2.COOH group derived from oxaloacetate (left in the formulae) being radioactive, in contrast to that derived from acetyl-CoA. The labelled citric acid is expected to form two different kinds of isocitric and α-oxoglutaric acids, namely

$$\underset{COOH}{\star COOH.\underset{|}{CH}.CH.CH_2.COOH} \quad \text{and} \quad \underset{COOH}{\star COOH.CH_2.\underset{|}{CH}.\overset{OH}{CH}.COOH}$$

$$\star COOH.CO.CH_2.CH_2.COOH \quad \text{and} \quad \star COOH.CH_2.CH_2.CO.COOH$$
$$+ CO_2 \qquad\qquad\qquad\qquad + CO_2$$

In the left-hand formula of α-oxoglutaric acid the radioactivity is located in the carboxyl group adjacent to the CO group, while

FINAL FORMULATION OF CITRIC ACID CYCLE

in the right-hand formula the radioactivity is located in the carboxyl group remote from the CO group. Thus the radioactivity is expected to be present in both carboxyl groups. Contrary to expectations, however, the radioactivity was detectable almost exclusively in the carboxyl group adjacent to the CO group derived from the oxaloacetic acid chain, and almost completely absent from the carboxyl carbon derived from acetyl-CoA. Thus it was established that the two .CH$_2$.COOH groups of citric acid do not react in the same manner and this was taken to prove that citric acid cannot be an intermediate in the cycle. Wood and his collaborators pointed out in 1941 that a minor modification of the cycle would meet the facts, namely the assumption (already mentioned in Chapter 9) that the condensation of oxaloacetic acid with pyruvic acid or a pyruvic acid derivative (this was before acetyl-CoA had been discovered) yields primarily aconitic acid which is then directly converted to isocitric acid, while the formation of citric acid is due to a minor side reaction

$$\begin{array}{c} \text{oxaloacetic acid} \\ \downarrow \quad \text{pyruvic acid} \\ \text{citric acid} \rightleftharpoons \text{aconitic acid} \\ \downarrow \\ \text{isocitric acid} \\ \downarrow \\ \alpha\text{-oxoglutaric acid} \end{array}$$

If the rate of the side reaction between citric and aconitic acids is slow compared with the rate of the other reactions, it is to be expected that the carbon atom derived from CO$_2$ appears predominantly in the carboxyl group of oxoglutarates adjacent to the CO group—as is the case.

This view was accepted for seven years. Then in 1948 Ogston revealed a fallacy in the assumption that citric acid *necessarily* behaves as a symmetrical molecule when it combines with an enzyme. Enzymes are asymmetrical molecules built from asymmetrical amino acids and, according to Ogston, an asymmetrical enzyme which attacks certain types of symmetrical compounds can distinguish between identical groups of a substrate provided that the substrate is rigidly attached to the surface of the enzyme and that there are differences in the catalytic properties of the various sites where binding between enzyme and substrate occurs. To appreciate Ogston's argument it is necessary to picture the substrate molecule in space. With citric acid as an example the treatment may be simplified by confining it to the central carbon atom only. In Fig. 7 this is pictured as being in the centre of the tetrahedron and attachment to the enzyme surface is taken to occur through the two .CH$_2$.COOH groups (represented by the symbol B) and the .COOH group (represented by C) in a plane below the central carbon atoms. There is only one position in which citric acid can be connected to the enzyme if a combination occurs at three points.

It was a matter of simplification to define the rigid combination of the substrate with the enzyme as a 'three-point attachment' because three points are the minimum required for

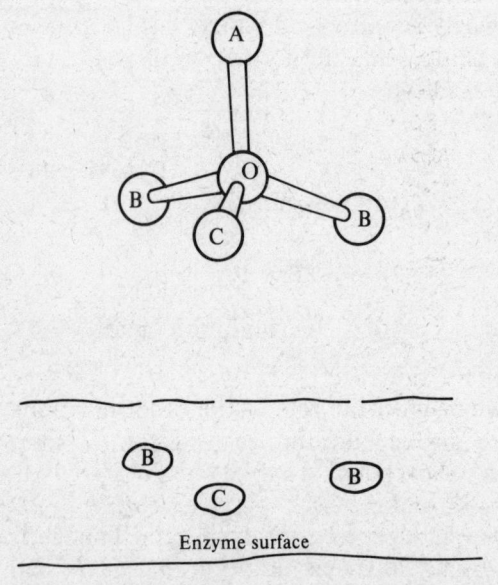

Fig. 7.   Three-point attachment of citric acid to enzyme surface.

O represents the central carbon atom of citric acid; A the –OH groups; B the two –CH$_2$.COOH groups; and C the –COOH groups. There is only one position in which citric acid can be connected into the enzyme if a combination occurs at three points.

FINAL FORMULATION OF CITRIC ACID CYCLE

rigidity, just as a table or chair requires at least three legs. In reality the substrate may also fit rigidly into a groove where it has many points of contact. Attachment at one point permits rotation of the attached molecule and thus gives scope for positioning in many planes; attachments at two points allows for rotation around one axis, which allows many alternative positions, but a three-point attachment is rigid.

Ogston also pointed out that his concept could account for the enzymic formation of optically active, that is, asymmetrical, substances from optically inactive (symmetrical) precursors. This is a frequent occurrence. Examples are the formation of L-malate from fumarate or L-lactate from pyruvate. Fumarate forms L-malate by the addition of the elements of water to the double bond:

$$\begin{array}{c} COOH \\ | \\ CH \\ \| \\ CH \\ | \\ COOH \end{array} + H_2O \rightarrow \begin{array}{c} COOH \\ | \\ CHOH \\ | \\ CH_2 \\ | \\ COOH \end{array}$$

If the fumarate is attached rigidly to the enzyme surface then the water can approach the double bond from one side only. This leads to the formation of either L or D malic acid. If both double bonds accept water equally, DL-malic acid is formed.

This work by Ogston was the start of the development of a large branch of stereochemistry.[20] As Bentley, reviewing the field in 1978 (1), put it, 'Ogston had uncovered the tip of an iceberg'. The subsequent work (2, 3, 4) was especially concerned with an analysis of the type of compound which can behave like citric acid. Thirty years after his discovery, Ogston commented on the question of how he came to make his discovery. He wrote (5):

My undergraduate training was pure chemistry, with a strong preference for physical chemistry. . . . Some time in 1948, when spending a spare hour looking through journals—mainly with a view to advising my pupils what to read—I came across a paper (by D. Shemin) which seemed to show that aminomalonic acid could not be an intermediate in the conversion of serine to glycine. I read it with interest and, initially, with consent. Suddenly, something happened. One moment I thought, 'That's neat'; the next, 'But it's wrong!'; the next, 'Citrate!'. It may have taken five seconds, perhaps less. A day or two later I sent off my letter to *Nature*.

Why did I leave it (almost) at that? Until I received, a few days after the publication of my note, an excited letter from Hans Krebs (at Sheffield) I had little notion of the size of the iceberg beneath that tip, regarding my idea as no more than an amusing piece of stereochemicology, and its consequences for asymmetric synthesis as merely an obvious corollary. And I was by then deeply involved with other lines of research which, if they have proved to be less sensational, have been satisfying in requiring far greater intellectual effort. 'Three-point attachment' was a gift, out of the blue, for which I have never felt able to claim much credit.

Yet I had, I suppose, unwittingly prepared myself for the three-dimensional visualisation that is required. As an undergraduate student I had enjoyed and I had been excellently taught the intricacies of stereochemistry.

Which confirms the wisdom of Pasteur's dictum, 'Le hasard ne favorise que les esprits preparés' (Chance favours only the prepared minds).

My letter dated 23 December 1948 to which Ogston refers, contained this passage, 'I am quite thrilled by your note in *Nature*. Now that you have pointed out the fallacy it is so obvious that one is surprised that it had been overlooked. Your theory seems to me a great step forward.'

In 1975 two scientists who took a leading part in developing the field, John Warcup Cornforth and Vladimir Prelog, jointly received Nobel Prizes. Cornforth's Nobel Lecture (6) began with a reference to the 'short but historic note by Alexander Ogston'.

15
THE NOBEL PRIZE
(1953)

To some, the news of a Nobel award may come out of the blue. In my own case there was plenty of premonitory rumbling.

I had never thought of the possibility of being in the running for the Prize until March 1951 when at a Ciba conference in London, Professor Einar Hammarsten, the Stockholm biochemist, engaged me in conversation and dropped unmistakable hints. Without mentioning the word 'Nobel' he skirted all round the subject of the Prize, telling me that much of his time during the summer was taken up in reading the literature and preparing reports. 'A very hard job', he said, 'but well paid'. He said he had read my papers again and again. I thanked him for his interest. Feeling somewhat embarrassed by his attentions, I tried to steer the conversation to other matters and said that I regretted never having been to Stockholm. He said immediately that he would try to arrange a visit for me, though, he added, the matter would not be all in his own hands.

At another Ciba conference a year later, in June 1952, Gerty Cori, herself a Nobel Laureate, whispered, 'I hope you will get the Prize this year'.

Four months later (a year before I received the Prize) a Swedish journalist who introduced himself on the telephone as Mr Bertil Askelöf, said that he had been instructed by his paper to interview me within the next few days. He wanted to be frank with me: I was a candidate for the Nobel Prize in Physiology and Medicine. I suggested to him that an interview should be postponed until an official announcement was made, that he would be wasting his time if his information was incorrect. To this he replied that this kind of information had never been incorrect on previous occasions, so, flushed and flattered, I agreed to an interview but refused his request to bring a photographer to take pictures of the family.

On the following Saturday afternoon, Mr Askelöf came to

Sheffield, and asked innumerable questions about my private life—the scientific aspects of my work would, he said, be dealt with by other people. He spoke in rather less confident tones than before and said that although his paper had information that I was a strong candidate, this did not mean that my nomination was a certainty. He stayed for about an hour and on departing warned me kindly, with reference to the prize money, that according to an old Swedish proverb, 'Fattiga förblir fattiga även om det regner guld' (The poor remain poor even if it rains gold).

After he had left I was still somewhat disbelieving but later came to the conclusion that his story must be serious, because I had heard of other instances where recipients had had first news of their Prize from a Swedish reporter. So Margaret and I began to prepare ourselves for 'all eventualities' but kept the approach from the press strictly to ourselves. Anticipating an interesting time ahead I started to keep a diary (for only a few days as it turned out!) and bought a scrapbook for press cuttings.

Mr Askelöf's enquiries were repeated later the same week by the London correspondent of *Dagensnyheter* and by the *Daily Express* who published that I was 'tipped' for the Prize. This item unleashed a flood of frantic enquiries from other journalists and an invasion of cameras. A photographer from the *Daily Mail* said he had not been told the reason for his mission, but had done a similar 'rush job' a few years previously when pictures of Professor Blackett were needed a few days before the announcement of 'a certain award'. The photographer could put two and two together: scientists (unlike film stars) hit the headlines only if they discovered a new atom bomb or received a very high award. He thought it unlikely that a biochemist had invented a new bomb.

At each fresh encounter, by telephone or face to face, I argued that the stories that were circulating were only rumours and that it would be wrong to publish unreliable information. Some I managed to convince, others not.

These were exciting hours but Margaret and I remained calm, and prepared for disappointment. Margaret said, 'This is what I would do instinctively'. So we carried on with our normal routines as far as possible, until the critical day, 23 October, when the announcement was expected in the evening.

THE NOBEL PRIZE (1953)

During that day, several reporters and photographers came or telephoned to the University or to my home. One member of the research team, R. E. Davies, asked me whether I had anything to tell the team—the laboratory was 'seething with rumours'. There was little peace during the evening. The *News Chronicle* insisted on taking pictures of the family; if I made them wait for the official announcement, they said, they would have to come back in the middle of the night. At 8.40 p.m. the *Daily Telegraph* congratulated me on the award. At 8.45 the *Daily Mail* congratulated me on the award. At 8.55 the *Yorkshire Post* said the Stockholm papers had tipped me for the award.

At 11.30 the *Daily Telegraph* apologized for their error 'due to a mistake by an otherwise trustworthy agency'. And so to sleep.

On the following morning, many papers reported the award to Dr Waksman and mentioned me as 'runner-up'. Although many world papers carried the correct story, many others printed the erroneous news that I had been awarded the Prize, and some that Waksman and I shared it. As a result, I received many letters of congratulation, to the later embarrassment of the writers and myself. Even more people congratulated me—and the members of my laboratory decided to give me a grand dinner—for having been 'runner-up'.

The awarding of the Prize jointly to Fritz Lipmann and myself a year later was heralded by similar experiences. The first request for a photograph came on 12 October from Hugo Theorell, the Stockholm biochemist, who wrote, 'I am trying to make a small photographic collection of outstanding biochemists for my Institute. Could you possibly send me a picture of yourself?' With the echoes of October 1952 still whispering in our ears, Margaret and I could not help but wonder if something else lay behind the request.

A week later we were back in the now almost familiar round of telephone calls and visits from press reporters and photographers, and on 20 October *The Scotsman*, *Sheffield Telegraph*, *Yorkshire Post*, *Daily Mail*, and many foreign newspapers published the unofficial news that Lipmann and I might receive the Prize. The announcement was expected on the evening of 22 October, so the next two days were busy.

We listened to the BBC news at nine o'clock on the 22nd and heard the announcement of the award, and half an hour later the official telegrams arrived from the Caroline Institute and the Swedish Ambassador.

Fritz Lipmann, with whom I shared the prize, was a friend of long standing and I had many memories of pleasant and stimulating times spent with him between 1927 and 1930 when we worked under the same roof in Berlin, Fritz with Otto Meyerhof and I with Warburg. Fritz had left Germany in 1930 for the Rockefeller Institute in New York and later went to Copenhagen, emigrating finally to the United States in 1939. Apart from seeing him at a meeting in Zürich in 1938, we had been only in written contact with each other since his departure from Dahlem. When we later met in Stockholm for the Nobel ceremony, his wife Freda dubbed us 'brothers-in-arms'.

Three of my fellow Laureates were not exactly strangers. As well as Fritz Lipmann, there was Hermann Staudinger, who was to receive the Chemistry Prize. Staudinger had been Chairman of the Department of Chemistry when I was at Freiburg and had been very helpful in getting recognition and appreciation of my work on urea synthesis. At Stockholm we were able to renew our acquaintance, and I met his wife, Magda, for the first time. She had been primarily a biologist but after her marriage she collaborated closely with her husband on the chemistry of macromolecules. After his death in 1965 she took a very active part in international work (within UNESCO) on the preservation of the biosphere. We have remained in friendly contact ever since our first encounter in Stockholm.

The Literature Prize had been rewarded to Winston Churchill who was, alas, unable to attend the presentation ceremony. He was then Prime Minister and was committed to attend a conference with President Eisenhower in Bermuda on 10 December, the day of the Nobel prizegiving. Lady Churchill, accompanied by her daughter, Mary Soames, represented him at the ceremony.

The only stranger to me was Frits Zernike, a Dutchman who received the Prize for Physics for his discovery of phase contrast microscopy.

On the morning after the announcement, the flag went up on the main University building at Sheffield as a token of

```
Charges to pay                POST      OFFICE IR0404   No.
    s.    d.                                            OFFICE STAMP
RECEIVED                       TELEGRAM                 SHEFFIELD
                 Prefix. Time handed in. Office of Origin and Service Instructions. Words
                     85        STOCKHOLM 64 22/10 2035   22 OCT 53
At
From   TS 0           TSO 1 326  GN L 487        To
By                                                By
```

PROFESSOR HANS ADOLF KREBS DEPARTMENT OF
BIOCHEMISTRY UNIVERSITY OF SHEFFIELD SHEFFIELD10 =
THE CAROLINE INSTITUTE HAS DECIDED TO AWARD THIS
YEARS NOBEL PRIZE IN PHYSIOLOGY AND MEDICINE WITH
ONE HALF TO YOU FOR YOUR DISCOVERY OF THE CITRIC
ACID CYCLE AND THE OTHER HALF TO PROFESSOE FRITZ
FALBERT LIPMANN FOR HIS DISCOVERY OF COENZYME A
AND ITS IMPORTANCE FOR THE INTERMEDIARY METABOLISM =

STEN FRIBERG RECTOR + +

CT SHEFFIELD10 +

```
Charges to pay                POST      OFFICE  IR0409
    s.    d.                                            No.
RECEIVED                       TELEGRAM                 OFFICE STAMP
                                                        SHEFFIELD
                     88                                 22 OCT 53
                      2
At
From  TS 0     F250     8.16 LANGHAM TS 71
By
```

= PROFESSOR HANS ADOLF KRABS A19 KINGSLEY PARK
AVENUE SHEFFIELD =
= ALLOW ME TO CONVEY IN THIS MANNER MY CORDIAL
CONGRATULATIONS TO THE HONOUR BESTOWED ON YOU BY
SHARING WITH PROFESSOR LIPMANN THE AWARD OF THE
NOBEL PRIZE FOR MEDICINE AND PHYSIOLOGT STOP IT
IS A SPECIAL PLEASURE FOR ME TO HAVE TO
CONGRATULATE FOR THE SECOND TIME THIS AUTUMN AS
NOBEL PRIZE WINNER A CITIZEN OF THIS COUNTRY =

GUNNAR HAEGGLOEF SWEDISH AMBASSADOR + (

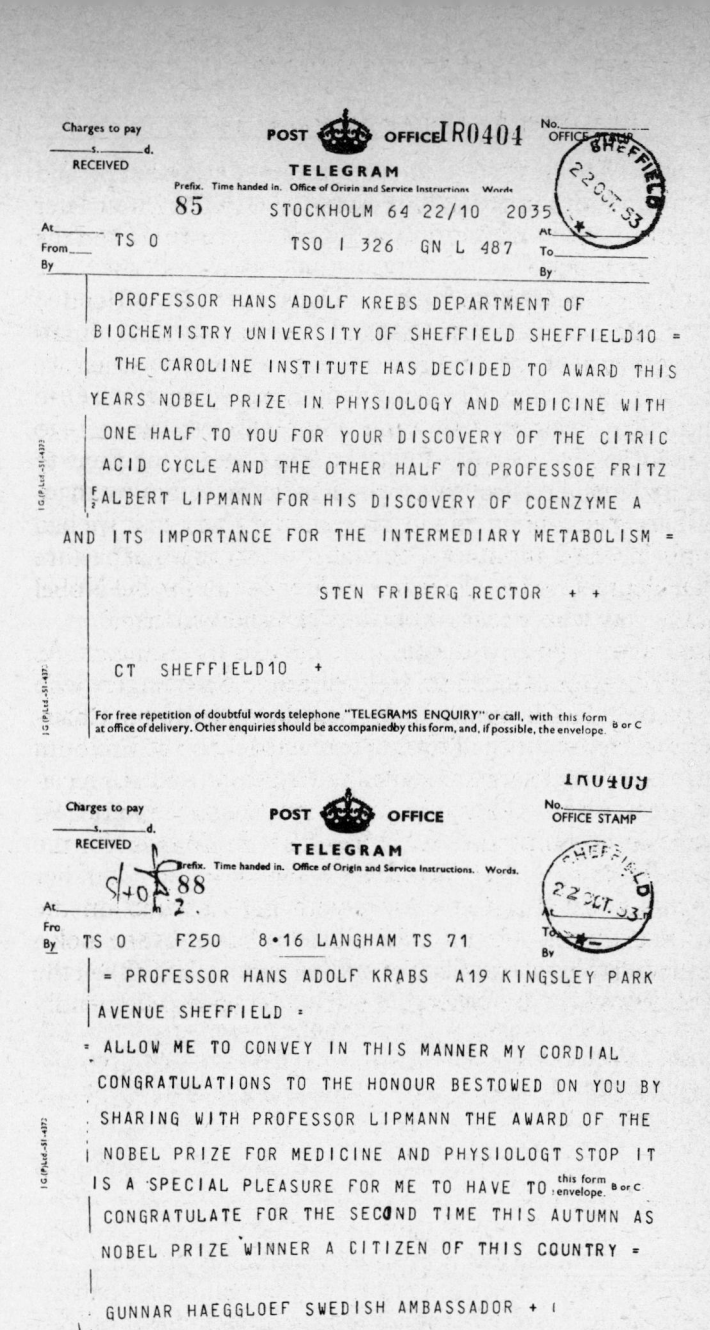

Fig. 8. Official telegrams from the Caroline Institute and the Swedish ambassador announcing the winning of the Nobel Prize, 1953.

celebration. When I arrived at the laboratory at nine o'clock I found my desk decorated with flowers and covered with letters and telegrams; within twenty-four hours I received a hundred and fifty messages of congratulation. Remembering the 'runner-up' dinner party which the Department had given me the previous year, I at once arranged another in return for the whole Department, which took place on 27 October.

Among the messages of congratulation were some which gave me particular pleasure: they came from long-lost friends and acquaintances. The period of Nazi rule in Europe and the war had severed many relationships which in normal circumstances would have been maintained. Thoughts of these lost friends had often been in my mind, but most of my attempts to re-establish contact had failed. The publicity of the Nobel award let them know where I was and many of them used the opportunity to get in touch with me.

The Faculty of Medicine at Freiburg, of which I was a member when Hitler came to power, sent a warm and touching letter which re-established formal contacts for the first time in twenty years. The letter was signed by Professor Zöllner, Dean of the time. After expressing cordial and sincere congratulations, it continued,

I find it difficult to express with words what I feel while writing these lines when I think back to the most saddening circumstances under which you had to leave your place of work as a young man in the early stages of your career as a university teacher.

How deeply were we all moved when your name appeared, together with that of Professor Staudinger, among the Nobel Laureates, and how proud could our Faculty have felt today if you had been one of our members.

I believe I am right in saying that the beginnings of your research career were in our Faculty, but were interrupted by the terrible times which have destroyed so many outstanding values. Perhaps it may be a conciliatory thought of your days at Freiburg if I enclose a copy of Professor Rehn's testimony, because his warm words were an honest expression of the views arrived at by his incorruptible judgement. This spirit of the Faculty has survived in spite of all the storms.

The Faculty would be exceedingly pleased if we could welcome you personally again in Freiburg and we would be grateful if on such an occasion you would lecture to us.

With the expression of a deep sense of attachment, I remain,
Yours very sincerely
Zöllner, Dean

Enclosed with the letter was the testimony written by Professor Rehn in 1932, which I had not seen before. (I have reproduced the text in Chapter 4.)

Zöllner's letter was followed a year later by an invitation to present the 1954 Aschoff Memorial Lecture.

The civic authorities in Sheffield also took note of the event, and I was touched to be presented by the City Council with an illuminated Resolution of Congratulations at a ceremony in Sheffield Town Hall in February 1954.

A rather different expression of recognition came from one of my elder son's schoolmates on the morning after the announcement; he met him with the cheerful greeting, 'Hello, Paul, I see your Dad's won the pools!'

The invitation to the Nobel ceremony and banquet includes one's family, so Margaret, and Paul, Helen, and John (now fourteen, eleven, and eight years of age) came to Stockholm with me. The press and their cameras were constant companions during our stay from the moment we arrived at Stockholm station on 7 December, and they interested themselves in every member of the family. John was asked about his favourite interests, and answered, 'Football'. I was pressed to say how I intended to spend the Prize money, about £6,000. At that time we had not made up our minds (eventually we bought a house in Oxford) but because I felt I must say something, I replied that I would buy myself a really good fountain pen; this great news made headlines in many newspapers! Early next morning, after a telephone call from the hotel's reception desk, requesting an urgent interview, two very serious looking gentlemen arrived in our hotel room. The senior of the two greeted us formally and announced with great solemnity that he wished to introduce me to a lifelong friend—a Montblanc fountain pen. He was the Swedish agent of the German manufacturers. There was a second pen for Margaret. A little later a journalist arrived with the gift of a football for John. (The boy has since broadened his range of interests and has become a behavioural zoologist.)

A visit to the British Embassy the day after our arrival briefed me on many technical aspects. The Ambassador, Mr Roger Stevens, gave me a great deal of background information and explained that because of the presence of Lady Churchill, he would not be able to spend as much time with me as he usually devoted to a British Laureate.

On the day before the ceremony there was an afternoon reception at the headquarters of the Nobel Foundation, where we met the other Laureates, the Officers of the Foundation and many of the leading Swedish scientists. On this occasion, Lady Churchill engaged me in a long conversation, asking me, obviously after having thought about it beforehand, what my feelings were, as a refugee, towards Germany and its people. We spoke about the views which I expressed in Chapter 13 of this book.

On the morning of the day of the prizegiving ceremony, 10 December, there was a thorough rehearsal of the proceedings, under the guidance of Officers of the Nobel Foundation.

The great Nobel prizegiving ceremony has often been described. A festive audience in evening dress assembles at 4.30 p.m. in the Stockholm Concert Hall. An orchestra plays music appropriate to the dignity of the occasion. The audience stands while the King and Queen enter with members of the Royal Family to take their places in the front row. Last to enter are the Laureates who take their seats on a dais facing the audience while the whole assembly, including the Royal Family, remains standing. It is the one and only occasion when the King is not the first to sit, when the Laureates are honoured in a manner reserved only for royalty, when learning and literature have precedence over noble blood. After a short laudatory citation to present each prizewinner, the King then hands to each Laureate the Nobel Medal and an illuminated document recording the award of their Prize.

The obverse of each medal carries the profile of Alfred Nobel. On the reverse (of the medal of Physiology and Medicine) the Spirit of Medicine, a book open on her lap, holds a basin to a rock spring to collect water to give to a sick girl. Around the margin is a paraphrased passage from Virgil's *Aeneid* (VI, 663), 'Inventas vitam juvat excoluisse per artes' (It is gratifying to have enriched life through discoveries).

THE NOBEL PRIZE (1953)

The gilded, leatherbound documents are individually designed for each Laureate. My own has a watercolour by Berta Svenson-Piehl of a lemon tree bearing many fruits, symbolizing the citric acid cycle; birds in the sky signify the spread of the beneficial knowledge through the world, and in the background is a scene of Hildesheim's market place and its most famous medieval building, the Hall of the Fleshers' Guild.

The Prizegiving is a most impressive ceremony; to be at its centre is a moving and unforgettable experience.

After the prizegiving a waiting car took us to the banquet at the City Hall, and from there we moved to another part of the building, the Blue Hall, where a ball had been arranged. There we were welcomed by university students who sang for us, a speech was given by the chairman of the Students' Association, and Professor Staudinger replied.

From the floor, wide steps led up to a spacious gallery where the members of the Royal Family, the families of the Laureates and the dinner guests were seated at tables, able to watch or take part in the dancing on the floor below. There was a charming expression of affection for Lady Churchill when a group of students in the hall burst into a spontaneous chorus of 'Oh, my darling Clementine'. Our children, Helen and John, were seated some distance away from us in the care of a special hostess, but after a while they dashed away to join us. As they passed King Gustav and Queen Louise, the King beckoned to them, and talked to them for a few moments. Having nothing else to offer, the King dipped a piece of sugar into his cup of coffee and popped it into Helen's mouth.

During the following days we had many other official and semi-official engagements. On 11 December I delivered my Nobel Lecture [158] before an audience of scientists specifically interested in the technical details of the work for which the prize had been awarded, and in the evening there was a dinner at the Royal Palace for the Laureates and many of the other leading participants in the ceremonies.

The Laureates received many invitations from Swedish scientists to visit their departments. During our stay, I visited the Caroline Institute and the University of Uppsala. On the evening of 13 December we took part in a student festival, the Lucia Festival, a light-hearted occasion with dancing. Here, in a mock ceremony, we were invested with the 'Order of the Frog'. There were no citations, no illuminated manuscript, just a green frog to hang around the neck.

We left Stockholm after nine delightful days, deeply touched by the friendly, enthusiastic, and generous reception we had had, not only from the Nobel Foundation and the scientific community, but also from the people of the city. Once home again, Margaret, wistfully but with some measure of relief, summed up our spell of limelight and lionization with the words, 'And now we are back to the obscurity whence we came'.

It did not work out quite like that. A little of the glory was to linger for a long time—sometimes in an embarrassing way. It is like having been stamped with a hallmark, a label of 'quality'. This would be very nice if it helped in obtaining grants for research support, but this I have not found to be the case. Grant-awarding bodies argue that although a Nobel Laureate has probably done good work in the past, this is no guarantee that he will do something worth while in the future. The time-consuming—and in my view far too elaborate—machinery of applying for grants is no less tiresome for a Nobel Laureate than for others.

The 'hallmark' is often used by organizers to boost the prestige of an event. I have felt this when invited to social functions and on being presented to distinguished guests of honour, not because I am Professor Krebs but because I am a Nobel Laureate. As James Franck (who received the Prize in 1925) warned me, 'You will often be invited as a "table decoration"'. One is asked to sponsor, if not to make donations to, innumerable good causes. Complete strangers ask for your comments and advice on all sorts of odd problems, assuming that Nobel Laureates can answer any question. One is pursued by autograph hunters, especially after having given a lecture. On one occasion a student asked me to sign no less than five cards for her.

'Why five?', I asked.
'I need the extra ones for swapping.'
'What do you get in exchange?'
'Oh, one really good movie star costs about five Nobel Laureates.'

THE NOBEL PRIZE (1953)

Students have often asked me, 'How does one become a Nobel Laureate?'[21] Because Nobel awards are to some extent a matter of good luck—they are too few to do justice to all who merit them—a more appropriate question would be, 'How does one attain excellence in science?' A methodical way of finding an answer is to study the history and characteristics of scientists of distinction. To do this one needs a criterion of distinction. So, despite what I have just said (and despite some personal embarrassment) I will use the Nobel award as a mark of distinction, for want of a better criterion.

When I asked myself how it came about that one day I found myself in Stockholm, I had not the slightest doubt that it was because I had had, in Otto Warburg, an outstanding teacher for four years at a critical stage of my development. This led me to wonder to what extent teachers had influenced other Nobel Laureates.

The sceptic might well suspect bias in favour of giving prizes to pupils of Laureates: in short, does nepotism play a part in the awards? The answer to this question is an emphatic 'No'. The high standing, in the eyes of the world, of Nobel awards is derived from the general recognition of the absolute integrity of the Nobel committees, and from the knowledge that these committees take a tremendous amount of trouble in finding the most worthy persons.

What, then, is it in particular that can be learned from teachers of special distinction? Above all, what they teach is high standards. We measure everything, including ourselves, by comparisons; and in the absence of someone with outstanding ability there is a risk that we easily come to believe that we are excellent and much better than the next man. Mediocre people may appear big to themselves (and to others) if they are surrounded by small circumstances. By the same token, big people feel dwarfed in the company of giants, and this is a most useful feeling. So what the giants of science teach us is to see ourselves modestly and not to overrate ourselves.

Warburg's teacher, Emil Fischer, was one of the outstanding chemists of his time. He was awarded a Nobel Prize in 1902 for his work on the chemical structure of sugars, the first of his long series of great achievements. Fischer in turn was a pupil and prolonged associate of another Nobel laureate, Adolf von

Baeyer, who received the Nobel Prize in 1905 for his discoveries in the field of the chemistry of dyestuffs, in particular for the synthesis of indigo.

Since Nobel awards began only in 1901 this criterion of excellence cannot be used for the assessment of excellence in the nineteenth century, but the following scientific 'genealogy' of earlier teachers and pupils shows that von Baeyer was a pupil of Kekulé (famous for his contributions to the structure of organic compounds, especially the ring structure of benzene), and that Kekulé was a pupil of Liebig (who laid the foundation of organic chemistry).

$$
\begin{array}{rl}
\text{Berthollet} & 1748-1822 \\
\downarrow & \\
\text{Gay-Lussac} & 1778-1850 \\
\downarrow & \\
\text{Liebig} & 1803-1873 \\
\downarrow & \\
\text{Kekulé} & 1829-1896 \\
\downarrow & \\
\text{von Baeyer} & 1835-1917 \\
\downarrow & \\
\text{E. Fischer} & 1852-1919 \\
\downarrow & \\
\text{Warburg} & 1883-1970 \\
\downarrow & \\
\text{Krebs} & 1900-
\end{array}
$$

Had there been Nobel awards at the time, Liebig and Kekulé would certainly have been Laureates. Liebig was a pupil of Gay-Lussac, discoverer of some of the fundamental laws of the behaviour of gases. Gay-Lussac was a product of the great French school of chemists, including in particular Berthollet, who pioneered in the concepts of combustion (abandoning the phlogiston theory in favour of the role of oxygen) and elucidated the chemistry of chlorine, ammonia, and hydrocyanic acid. One of Berthollet's teachers was Lavoisier.

In every case the association between teacher and pupil was close and prolonged, extending to what we would now call post-graduate and post-doctoral levels. It was not merely a matter of attending a course of lectures but of researching together over a period of years.

My scientific 'genealogy' drives home the point that, in many instances, distinction breeds distinction or, in other words, distinction develops if nurtured by distinction. This is further borne out very forcibly by a consideration of a more extended family tree of scientists which summarizes the genealogy of the Nobel laureates descended from von Baeyer, the pupil of Liebig, and includes seventeen names. Outstanding discoveries can be associated with every one.

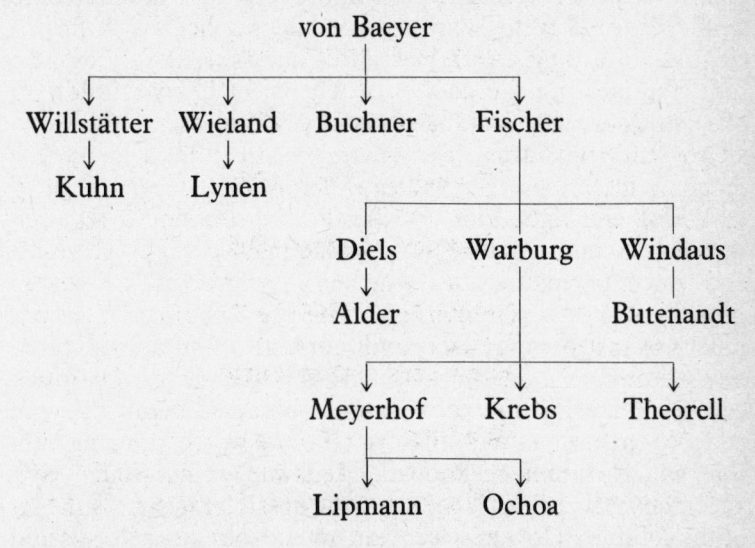

```
                           von Baeyer
           ┌──────────┬────────┬──────────┐
           ↓          ↓        ↓          ↓
      Willstätter  Wieland  Buchner    Fischer
           ↓          ↓              ┌─────┼──────┐
         Kuhn       Lynen            ↓     ↓      ↓
                                   Diels Warburg Windaus
                                     ↓            │
                                   Alder       Butenandt
                                          ┌─────┼──────┐
                                          ↓     ↓      ↓
                                      Meyerhof Krebs Theorell
                                          ↓     ↓
                                       Lipmann Ochoa
```

A similar family tree could be assembled for the physical sciences (see also Harriet Zuckerman) (1).

Peter Medawar, (in an unpublished speech) commenting on science family trees, made the point that each 'generation' of researchers differed from its teachers, often very radically, in fields of research and manner of approach. Pupils had not followed exactly the footsteps of their teachers. But they had learned the principles involved in exploring new regions, so they themselves became pioneers. The new orientation of each generation has been responsible for an extraordinarily rapid rate of progress in the sciences. Max Delbrück expressed the same idea in a different form: 'If a scientist at the age of 50 still understands all his pupils, he never had a really good student'. Delbrück was himself over 50 when he said this.

It is attitudes, even more than knowledge, that are conveyed by the distinguished teacher. Technical skills can be learned from many teachers and, like a modicum of intelligence, are, of course, prerequisites for successful research. What is critical is the use of skills, how to assess their potentialities and their limitations; how to improve, to rejuvenate, to supplement them. But perhaps the most important element of attitude is humility, because from it flows a self-critical mind and the continuous effort to learn and to improve. If I try to summarize what I learned from Warburg I would say he was to me an example of asking the right kind of question, of forging new tools for tackling the chosen problems, of being ruthless in self-criticism and of taking pains in verifying facts, of expressing results and ideas clearly and concisely and of altogether focusing his life on true values.

Asking the right kind of question in choosing a research problem means avoiding those which may give a quick result and concentrating on those which are really worth while tackling. Paul Weiss (3) remarked: 'The primary aim of research must not just be more facts and more facts, but more facts of strategic value'. Goethe (4) expressed the same idea much earlier: 'Progress in research is much hindered because people concern themselves with that which is not worth knowing, and that which cannot be known'. Medawar (5) has stated very succinctly: 'If politics is the art of the possible, science is the art of the soluble'. How to select worthwhile soluble problems and how to create the tools required to achieve a solution is something that scientists learn from the great figures in science rather than from books.[22]

Association with a leading teacher almost automatically brings about close association with outstanding contemporaries of the pupil because great teachers tend to attract good people. Students at all levels learn as much from their fellow students as from their seniors and this was certainly true in my own case. Warburg's laboratory at Dahlem, where I served my apprenticeship, was surrounded by other centres of distinction, as I described in Chapter 3.

There are many other examples of such centres of excellence

and breeding grounds of scientists. Cambridge, for example, was a centre of excellence in physiology and biochemistry in the early decades of this country because Foster, Langley, Hopkins, Barcroft, and Adrian were each surrounded by a group of enthusiastic young people of great ability. Cambridge, of course, at the same time was also a centre of excellence in physics, thanks to J. J. Thomson and Rutherford.

There have, of course, also been 'self-made' Laureates. The great majority of these, however, though they had no single outstanding teacher, had around them many exceptionally gifted colleagues, who provided a highly stimulating and congenial environment. One of the great exceptions—in being entirely self-made—was Albert Einstein, who produced his early, great contributions in almost complete isolation. But here, too, there was something special and unusual about the environment: he had plenty of time for his research because his routine work at the Zurich Patent Office allowed him much free time.

In brief, then, to answer the students' questions on how to attain distinction in scientific research, you must

- get a post-doctoral Fellowship—which will give you the time and opportunity to test yourself;
- attach yourself to a centre of excellence;
- work hard and make the fullest use of the time and facilities the Fellowship affords;
- from time to time, search your heart critically, with the help of objective critics, to find out whether you really possess the right mixture of qualities—urge, commitment, imagination, humility—which are the roots of creativity in science.

16
ACADEMIC INVITATIONS
(1946–1954)

In 1946 I received my first approach about my possible interest in a professorship at another university. It came in a letter from the Dean of the Faculty of Medicine at the University of Basle asking if I would consider an offer of the Chair of Physiological Chemistry. But I had no wish to leave England, where I felt completely at home. Nine months later, in June 1947, I received an offer from Alexander Haddow, on behalf of the University of London, of a Chair of Biochemistry tenable at the Royal Cancer Hospital. Again I preferred to stay at Sheffield. I was more attracted to move when, in May 1951, I received a letter from George Wald, Professor of Biology at Harvard University, telling me that Harvard's Administration had decided to explore the possibility of appointing a small number of new Professors in the Department of Biology in order to bring a few men of high distinction into the Department.

Wald asked if I would care to be considered for such an appointment. I replied that I was very much interested but that I would need detailed information, and I raised a number of questions which Harvard decided could only be resolved in personal discussions. They invited me to spend a week there as a Visiting Lecturer, to give two or three lectures, and discuss the projected professorship. I accepted the invitation and visited Harvard from 16 October.

There proved to be many attractive aspects of a move to Harvard. I would have much more space, and facilities were good; I liked the 'multi-professor Department' concept where the administrative work of a Department was shared, and its 'Chairmanship' held by each professor in turn for a period of three to five years; also, I was in agreement with Harvard on the envisaged teaching requirements and programme. My main worry was whether, at Harvard, I could build up something equivalent to the very effective team of young but experienced 'independent' workers I had at Sheffield, although I had obtained Harvard's agreement in principle to a transfer of technicians and one instructor on a temporary basis.

At the end of my stay I was formally offered the post by Mr Buck, the Provost of Harvard. I replied that I greatly appreciated the offer but that I should have to think matters over. Mr Buck asked me to write from England stating my views and terms of acceptance; if they could not be met, discussions could be re-opened.

From Harvard I went to New York to discuss the proposition with Gerard Pomerat and Warren Weaver of the Rockefeller Foundation. Weaver told me that the performance of the Biology Department at Harvard had been rather disappointing when one compared output with outlay. He was equivocal about my moving. He would welcome the strengthening of biology at Harvard but would regret the disruption at Sheffield. He agreed that a set-up of the kind I visualized was essential and would not be against the principles of United States universities in general or of Harvard in particular. Back in England, the question caused me much heart-searching because Harvard offered a stimulating environment and splendid opportunities, and some of my circumstances at Sheffield were far from ideal. While the financial support of my team was fully adequate thanks to the Medical Research Council, my scope was severely restricted by lack of space both for research and for teaching. Pondering on the invitation, I decided to express my views to the Sheffield University authorities in a memorandum which I submitted to the university authorities in November 1951. In it I tried to assess the status of biochemistry at Sheffield from my point of view—admittedly personal and biased, but relevant not only to my own tenure of the Chair but also to that of a successor.

At that time about two-thirds to three-quarters of the total financial requirements of the Department came from outside bodies, such as the Medical Research Council, the Rockefeller Foundation, and other grant-awarding organizations. The Department enjoyed an international reputation as one of the foremost centres of biochemistry and attracted a large number of post-graduate students and senior visitors from all over the world.

In my memorandum, I made it clear that ever since its inception in 1938 the Department had suffered from inadequate accommodation. In 1947 space became even more cramped when the Department, having been invited to do so, accepted the additional responsibility of teaching about a hundred undergraduate students. Adding to this was the phenomenal progress of general biochemical techniques, especially the advent of radioactive isotopes in 1948, which necessitated additional equipment if the Department were to maintain its place in the forefront of research. Money was available to buy the equipment, but there was nowhere to put it.

The Department had no teaching laboratory and no lecture theatre; all teaching had to be done on premises shared with other Departments; no member of the staff, not excepting myself, had a private room, nor was there a general staff room which could be used as a private room for confidential interviews.

The memorandum ended:

The scant space given to the Department (1/6th of the space available to Chemistry and 1/3rd of that available each to Zoology and Botany) suggests that the general importance of biochemistry as an academic subject has not been appreciated by the Sites and Building Committee (the body, as far as I know, which is responsible for space allocation).

In all biological subjects, including medicine and agriculture, biochemical ideas and techniques are becoming increasingly important. One might go as far as to say that *all* major developments in biological subjects have biochemical aspects. A biochemistry department is therefore of importance to all departments in the University concerned with the study of living things. A manifestation of the increasing importance of biochemistry is the demand for biochemically trained scientists. The Department annually receives about 20 requests (almost all vain) for biochemists trained in the branches of the subject pursued in the Department.

I hope that the preceding remarks bring home my contention that the Department of Biochemistry has not had a square deal from the Sites and Building Committee. I gladly state that I have no criticism to make on any matter other than space allocation.

There are several grounds on which one can base a choice of post, such as salary considerations, local amenities, the status of the institutions, family considerations and the facilities offered for doing one's job. In my own case the last point, in its widest sense, is the overriding one and given adequate facilities I would choose to remain at Sheffield. The argument of the memorandum is, to sum up, that the present working facilities are grossly inadequate but that by devoting a very modest fraction—a few percent—of the University's immediate expansion programme to Biochemistry, the facilities could soon be made reasonably adequate.

I shall be grateful if the University Authorities will comment on the points raised in the memorandum and clarify the situation by making a statement of the University's short-term and long-term policy on the development of Biochemistry.

In response to this memorandum the University offered to provide additional space by converting two rooms in the adjacent Scala Cinema which had been acquired by the University. These rooms, which had served as a restaurant and dance hall, each had a floor area of about a thousand square feet. Both had gas, water, and electricity and could relatively easily be adapted as laboratories. This additional space made it possible to enlarge the teaching commitments of the Biochemistry Department by instituting an Honours School of Biochemistry.

In considering the invitation from Harvard, two things were uppermost in my mind. First I had a strong sense of attachment to my team. I felt that without them, my effectiveness in research would be greatly diminished and I was uncertain whether it would be possible to build up a similar group in the setting of Harvard. I was loath to risk breaking up a team which had so much potential.

Second, there were financial considerations, arising from the currency restrictions in force at that time. Of my total assets in Britain (consisting mainly of our house) no more than £500 per year could be transferred to the United States, and then for the first four years only. Personal effects and household goods were included in this sum. Also during the first four years, assets remaining in Britain could not be used to buy goods to be sent or taken out of the country; they could be used for holidays in Britain, but not to pay transatlantic fares; purchases in Britain were limited to £50 per year. At the end of the four years, all assets would be frozen, which meant that they could not be spent even in Britain; only accrued interest could be used or remitted abroad. The restrictions also applied to all insurance policies which matured and to legacies.

Thus we would have had to start our new lives in the United States with a mortgage and bank loans, without resources for emergencies. Margaret and I felt that this could be a great worry to us, having in mind especially the security and care of our three children. So I declined Harvard's offer.

This was the third time in my career that my circumstances were improved by pressure from outside. As already mentioned, in 1938 the Rockefeller Foundation had made their contribution to my work conditional upon the setting-up of a Department of Biochemistry under my headship, and in 1945 the Medical Research Council offered to support my research by establishing a Unit and thus guaranteeing longer-term financial support, conditional upon my status and salary being raised to that of a full professor, with the Medical Research Council contributing half of that salary.

I must mention that at the time I was discussing the Harvard offer with Sheffield University Authorities, there was general concern in Britain about the numbers of scientists leaving the country (the 'Brain Drain'). This prompted both the Medical Research Council and the Agricultural Research Council (on one of whose main committees I had served for many years) to make serious efforts—most gratifying to me—to provide me with adequate space in one of their research institutes. The Medical Research Council explored possibilities at the 'Serum Institute' at Carshalton and the Agricultural Research Council at Babraham, near Cambridge. But Sheffield University solved my problem and my preference was indeed to work in a university rather than in an isolated research institute. I was anxious to stay in contact with other departments and other faculties, and also with students.

Because, by the early 1950s, biochemistry had become a subject of increasing importance to every branch of biology, we in the Biochemistry Department at Sheffield felt that the time had come to introduce more intensive undergraduate teaching, in the form of an Honours School of biochemistry. Until then, undergraduate teaching had been restricted to lectures and practical classes for medical students plus a few lectures in the Honours Schools of chemistry and physiology.

ACADEMIC INVITATIONS (1946–1954)

Biologically oriented colleagues in the Departments of Zoology, Botany, Physiology, Pathology, Microbiology, and Medicine were all in favour of our proposal to introduce an Honours School in biochemistry, but some in the Department of Chemistry—a subject very close to biochemistry—were against it. They defended the view that biochemistry should be taught intensively only at the post-graduate stage, after a thorough grounding in pure chemistry. My colleagues and I held that chemistry and biology should be taught in parallel from the start because the ways of thinking are different in chemistry and biology: chemists think in terms of the properties of atoms and molecules; biologists think in terms of the functional operation of living things and so are interested in the role that chemical substances play in the living organism. Teaching experience shows that it is advantageous to introduce early in the student's training the study of general biology and to inculcate the ways of biological thinking. Otherwise many cannot get away from the chemical way of thinking and have difficulty in acquiring a 'feel' for the characteristic nature of biological systems, especially of the one-ness of a living creature in which all parts form the whole and contribute to its survival.

The disagreements at Sheffield led to a confrontation at a meeting of the Board of the Faculty of Science, but after a vigorous debate the Board accepted my proposal to introduce an Honours school. This welcomed its first students in 1952, and has flourished ever since.

I must add that my main opponent on the Board, Professor R. D. Haworth (who had opened his speech with the words, 'I have the honour to present the views of the opposition') after his defeat gave the biochemistry school his full support.

During the Easter vacation of 1954 I spent some nine weeks in the United States—a visit which had been arranged before the announcement of the Nobel award. I had been asked to deliver the Herter Lecture Series at the Johns Hopkins Medical School at Baltimore and had also received invitations from the Universities of Chicago and Wisconsin to spend a month at each as a Visiting Professor.

During my stay in the States I received two letters from Oxford, one from Kits von Heyningen of the Pathology Department (later Master of St Cross College), and one from Hugh Sinclair, who was at that time in charge of the Laboratory of Human Nutrition. Both drew my attention to the announcement in the *Oxford University Gazette* that the Whitley Professorship of Biochemistry would become vacant on 30 September 1954 through the resignation of Professor Sir Rudolph Peters; the electors invited applications by 29 April 1954. The letters encouraged me to apply, telling me that there were many people in Oxford who would welcome my coming.

I replied on 12 April:

Dear Sinclair,
Your letter reaches me during a visit to the United States. I appreciate very much your kind encouragement.

A copy of the advertisement was sent to me a few weeks ago and I have been thinking matters over for some time. I am of course very much attracted but my feeling is that I am not really free to apply without having first discussed the fate of my Unit with the Medical Research Council and I cannot just try to walk out.

These problems can hardly be dealt with by correspondence and I therefore do not find it possible to cope with the matter before I return. I am due to be back in Sheffield on 11 May. I noted in the advertisement that the Electors are not necessarily restricted in choice to those who apply. If they wish to consider me I shall be available immediately after my return.

With kindest regards and many thanks.

Sinclair wrote in reply that he had passed on my letter to one of the electors, Sir Edward Mellanby. He explained his action by saying, 'University Electoral Boards are curious things[23] and now and then they reach conclusions on entirely inadequate evidence. So Mellanby, being an experienced and important Elector, should know the position'. He attached a copy of his letter to Mellanby, which included the following passages:

When you paid your visit here we had a word about the forthcoming election to the Chair of Biochemistry. Since I have myself served on three electoral Boards here and since you have served on a great many more, it is common knowledge to us that it is not unusual to be lobbied, however annoying this may be. As a Biochemist I regard the appointment here as being an enormously important one both for the University and for the subject in general. I feel, as I mentioned to you, that Krebs is the outstandingly obvious candidate and I have ventured privately, since I am fortunate in knowing him, to make sure that he is interested. His letter in reply, of which I venture to enclose a copy, is typical . . . My main object in writing, therefore, is to make sure that you know that his failure to apply before 29 April does not mean that he is not interested.

On my return to Sheffield on 11 May I found the following letter from the Registrar of Oxford University.

8 May 1954

Dear Professor Krebs,

At a meeting of the electors to the Whitley Professorship of Biochemistry in this University, which was held this morning, I was directed to offer you the Professorship and to ask you whether you would be prepared to accept.

I enclose a copy of the particulars concerning the Professorship, which were published when it was advertised. I should be happy to give you any further information which you desire.

Yours very truly,
(signed) Douglas Veale

This letter of course pleased me very much and I showed it at once to my senior colleague, John Bacon, whose immediate reaction was, 'You must take this invitation very seriously'. John knew that I was very happy at Sheffield and deeply attached to my team, and he was therefore concerned that I might not give the offer due consideration. He explained that opportunities for promoting biochemistry would be very much greater at Oxford than at Sheffield; Oxford and Cambridge attracted a greater proportion of first-rate undergraduate and post-graduate students than other British universities. He was convinced that I would have much greater scope. I took his encouragement to heart and thought about the offer very seriously indeed.

On the following day, 12 May, I telephoned the Secretary of the Medical Research Council, Sir Harold Himsworth, and inquired whether a transfer of the Unit from Sheffield to Oxford would be feasible from the Council's point of view. He replied that he was prepared for this question. As the Unit at Sheffield was a personal arrangement which would cease in any case if or when I left the University, the Council would have no objection to its transfer. Whether or not the transfer would be practicable would depend upon the co-operation of Oxford.

Later in the day I telephoned Mr Veale, I thanked him for his letter and expressed my pleasure at the invitation. I asked whether there was a possibility of transferring my MRC Unit to Oxford. Veale replied that he had been prepared for this question and could say that the Unit would be welcome provided that physical space could be made available. We agreed that I would visit Oxford to discuss these matters in detail and Veale later wrote to inform me that he had arranged meetings for me on 18 May with the Vice-Chancellor, the Registrar and also with Professor Peters. The Vice-Chancellor at the time was Sir Maurice Bowra and the meeting was to be held at his rooms in Wadham College. A few days later I received a telephone message from Peters, suggesting that I might meet him at his home before seeing the Vice-Chancellor and the Registrar.

The discussions at Oxford were satisfactory and fruitful. Peters gave me much valuable information about the Department and about the impossibility of accommodating the Unit without a further allocation of space. The tone of the meeting with the Vice-Chancellor was set by Bowra's opening remarks: 'We want you here at Oxford, and we want you very much. We were expecting that you would require extra space for your Unit and we are keen to find it.' The meeting was brief and entirely to the point. All my questions were sympathetically answered. The University Surveyor was immediately instructed to explore the possibility of erecting two huts on a site adjoining the Biochemistry Laboratory, and a few days later I was informed that some two thousand square feet of additional laboratory area would be provided. The work could not, of course, be completed by 1 October 1954, the date when my appointment should start. The problem of having space available on 1 October was solved by the generous help of the Professor of Chemistry, Sir Robert Robinson. It had been planned to transfer the teaching of basic chemistry to biochemists, physiologists, and medical students from the Department of Organic Chemistry to the Department of Biochemistry on 1 October 1954. On the suggestion of Kits von Heyningen, I asked the Vice-Chancellor if this transfer could be postponed for one year; he conveyed this suggestion to Robinson, who readily agreed. This made a very large laboratory available to the Unit.

It was also helpful to me that I had the opportunity of

ACADEMIC INVITATIONS (1946-1954)

discussing the question of the continued support of my research by the Rockefeller Foundation with Dr Gerard Pomerat, of the Foundation. Dr Pomerat saw me in Sheffield on 14 May, and we met again in Oxford after my meeting with the Vice-Chancellor. He explained that, in the event of my leaving Sheffield University, the University should ask the Foundation to be relieved of the grant, since it was made for my personal research. Once established at Oxford, I could re-apply for the grant if I found conditions similar to those in Sheffield. If they were, in fact, identical, then I could make an identical request and there was no reason why the Foundation should not treat it in an identical way. He went on to say that it was not the custom of the Foundation to make grants for the first year—the 'honeymoon period'—in a new post; any special requirements for this period should be met by the University. In fact, they were, and the Oxford Registrar told me that I should make all my requests during the first year, which was indeed a honeymoon period during which I would be well treated. Later I should not expect too much from the University because there was always keen competition for funds.

I asked Pomerat for advice on accepting the offer and he replied that, as an officer of the Foundation, he was not entitled to give advice on matters such as accepting an invitation to a new post, but it was his personal opinion that he would consider it right to accept. He did not know of any reason against it but of many in favour.

A major concern to me was the willingness of the Medical Research Council Unit to keep the team together by moving it with me to Oxford. One of the stumbling-blocks for those members who were married was the cost and availability of housing at Oxford. This was especially true for the technicians. Two of them made an exploratory visit which revealed that unfurnished rented accommodation was extremely scarce. House purchase was beyond the means of the junior members. But the Medical Research Council's allowances towards removal expenses and house buying overcame the most serious difficulties. By the middle of June agreement had been reached that all but one of the key members (Bob Davies, David Hughes, Walter Bartley, Donald Exley, Kenneth Burton, Hans Kornberg, Leonard Eggleston, Reginald Hems and Alan

Renshaw), would move to Oxford. A few visiting scientists and research students were to accompany us. Derek Williamson decided to go his own way and joined the Courtauld Institute at Middlesex Hospital to work in the laboratory of Professor Frank Dickens. (As I have already mentioned, he rejoined the team six years later.) So, apart from two young technicians who lived with their parents in Sheffield and decided to stay behind, the Unit was, to my great satisfaction, to move without serious injury.

During June I travelled several times to Oxford, sometimes with my senior colleagues, to discuss details of our requirements. I met members of the Oxford Department, in particular Cyril Carter, Donald Woods, Rupert Cecil, Lloyd Stocken, Sandy Ogston, Kits van Heyningen, and Janet Vaughan.

During one of these visits I asked Rudolph Peters for his advice and for any general comments about the work I would be taking on. He made two points. He confessed that in his view, I might not be the right person for the job. As he saw it, the most important job waiting to be done at Oxford was to bring science and the humanities closer together, especially at the college level—as he put it, 'To sell science to the non-scientific academics.' 'Oxford', he continued, 'is governed and controlled by the thirty or so Colleges and in each College, the non-scientists are in the majority.' To add to this, every major board and committee had a non-scientific majority. To help science achieve appropriate recognition would, he thought, be too difficult for a newcomer; only a don with a firm footing in the colleges and with experience in their mode of working—such as Ogston or Woods—would be likely to succeed.

His second point was that, 'At Oxford the Professors are a suppressed class', always outnumbered on committees; the power and decision-making remained with the college dons. This was echoed by a passage in a (then) newly published book by Douglas Veale (1), in which he said, 'It is perhaps worth noting that in Oxford a professorship is not a grade in an administrative hierarchy as well as being the summit of an academic hierarchy... In the administrative sphere there is, between the individual and his faculty, no intermediate machine in which he can become entangled. Indeed, there can be few institutions in the world in which an official position deprives its holder of so much power'.

These comments were all very useful to me. I can see, in retrospect, that Peters was right to stress the importance of the recognition of science as an equal partner with the humanities, but his reservations about my effectiveness to help bring this about were not quite so justified. Within three years I became a member of the Hebdomadal Council (the main governing body) whose candidates are elected by secret ballot of all senior members of the University.

Soon after my arrival—and without any intervention on my part—a few colleges began to elect scientists to their Headships, which till then had been a rare occurrence indeed. They were Magdalen, Queen's, Hertford, Pembroke, St Cross, Merton, Trinity, and New College. This was the consequence of a general slow development—the time had just become ripe. There had been an increasing recognition by the non-scientists of science as a 'respectable' academic exercise. Old prejudices faded as more and more people came to appreciate the nature and objectives of science and its importance to national life. As a member of the Hebdomadal Council, on which I was to serve for five years, I was never aware of any 'anti-science' attitudes in my non-science colleagues.

On 21 June I learned from the Registrar that Council had agreed to all my proposals and on 23 June I was able to confirm that, 'I am glad to accept the election to the Whitley Professorship from 1 October 1954'.

In my discussions with the Vice-Chancellor I mentioned that the salary offered (£2300 per annum) was £200 less than I was receiving at Sheffield. At first the Vice-Chancellor thought the difference could be made up but later had to tell me that it could not. He consoled me, however, by saying that I would most likely earn additional fees by examining and that I would be entitled to free dinners in my college. I was not unduly worried—although some time later I let him know that he had grossly overestimated the size of my appetite. It was of much greater importance to me that the University had agreed to find £20 000 for the building of the two temporary laboratories.

I was now ready to tender my resignation from the Sheffield Chair. In my letter to the Vice-Chancellor I said, 'I need hardly tell you that it is with mixed feelings that I sever my links with Sheffield University and my colleagues, after many years of happy and fruitful association. I shall always be grateful for the opportunities that Sheffield University gave me.'

17
OXFORD
(1954–1967)

Oxford is in many ways a splendid place for academics, one of the finest in the world. The dominant atmosphere is one of tolerance, mutual helpfulness, respect, support, and the appreciation of academic endeavour without fuss, adulation, or self-adulation. It is a striking fact that, despite its ups and downs, Oxford has retained over many centuries the reputation of being one of the foremost world centres in education and in the furthering of scholarship. At times there have been equally distinguished, or even more distinguished, academic centres, such as Bologna in the middle ages, but no other except Cambridge has managed to remain young in spirit and in the forefront of learning for over some six hundred years.

What has kept Oxford in the forefront has been its broad basis—the many (now thirty-five) semi-independent colleges competing for excellence, both in the selection of students and in academic and sporting achievements—and that it has had the capacity to adapt itself to changing times on the basis of honest self-analysis and self-criticism. Oxford, and of course Cambridge, have succeeded more than most institutions in following St Paul's command: 'Try all. Hold fast to that which is good.' If other places deteriorated it was because they neglected this source of strength and because they suffered from corrupting influences which placed the self-interest of the teachers before the interests of the institution. So they became out of touch and out of date. Oxford has shown a special skill in combining the preservation of old, well-tried traditions with the preservation of a freshness of approach to the problems of the day. Not that Oxford ever was—or is—immune to eroding influences. The struggle between the reactionary (that is, the retrograde) and the progressive elements is ever present. I myself became deeply involved in debates on shortcomings which I felt ought to be remedied. Some of these battles consumed a great deal of time and energy and I sometimes wondered whether I should allow myself to be drawn into controversial issues. But I do not now regret that I did. On the whole, though by no means always, the progressive elements have prevailed.

One of Oxford's great assets is the attraction it exerts on the very best students, not only from Britain but, at the postgraduate level, from all over the world. While the average Oxford students may be no better than those in the newer British universities, as my experience at Sheffield suggests, the number of exceptionally gifted students is much greater at Oxford. Not only do they become the outstanding postgraduate workers, they raise the level of the whole class; they make the teacher feel that his efforts are really worth while; and they set a standard and an example to their fellows, because students learn as much from each other as from their teachers, technically, intellectually, and in their attitudes towards work and academic values.

Another very special attraction of Oxford is the college system.[24] For the professorial Fellows in the sciences the colleges are not a place of teaching; they are relatively small and close-knit social centres where one rubs shoulders with academics from all areas of learning and meets them informally at meals and social gatherings. I found this regular, casual contact with historians, classicists, lawyers, linguists, economists, and philosophers, as well as fellow-scientists, extraordinarily pleasant and profitable. It was a source of continual education. Nowhere else, I believe, is there less fragmentation into faculty and subject than at Oxford and Cambridge, thanks to the college system which brings people from all subjects together. While very much alive and up-to-date, the colleges reflect a continuity of tradition; their ancient buildings, their historical associations, their treasures of books and *objets d'art* preserving the civilization of previous centuries, telling those who are perceptive that (as the Hildesheim house inscription reminded me in my youth), 'Our forebears were no fools either'.

Then, too, it is a beautiful city with charming environs. The nearness of London and, through London's airports, the nearness of the whole world, adds to Oxford's attractions.

So, Oxford provides a magnificent background for research

work and I was extremely happy there. Despite some frustrating and exasperating experiences in the area of administration, never for a moment did I regret the move. I was fortunate to be in a position where, thanks to the Medical Research Council and other grant-awarding bodies, I was financially more or less free from budgetary restrictions on the part of the University and I was free, through my Medical Research Council Unit, to recruit my own research staff.

It was of great help to me that there were several people at Oxford with whom I had had earlier contacts. In the Department of Biochemistry there were, apart from Peters (with whom I had collaborated closely during the war years on vitamin C deficiency), Sandy Ogston, who had collaborated with the Sheffield team in the late 1940s, and Donald Woods whom I knew from my Cambridge days and whose Ph.D. work I had partly supervised. Percy O'Brien and Hugh Sinclair had frequently been to Sheffield to work on vitamin deficiencies on human volunteers. H. W. (Tommy) Thompson (later Sir Harold Thompson) I had met regularly in Dahlem in 1929–30 when he spent a year as a research student with Fritz Haber. Kits van Heyningen in the Pathology Department had been my collaborator at Cambridge. George Pickering, who joined Oxford University as Regius Professor of Medicine a year after my arrival, had been a friend ever since we had been introduced in Freiburg in 1932. Then there was my very old friend Hermann (Hugh) Blaschko, with whom I had been a student at Freiburg in 1919. Hugh and Donald Woods were of special value to me, giving me the benefit of their long experience of Oxford life and personalities.

I found the Department of Biochemistry in excellent shape. In the course of thirty-one years my predecessor had built up from scratch a high reputation and an effective teaching and research team. There was no dead wood in his Department. He had also prepared the way for further developments, especially in the case of the Honours School of Biochemistry and for expanding the Departmental buildings. The Honours School had been introduced in 1948 with two entrants. When I retired in 1967 the entry was over forty—an indication of the increasing recognition of the importance of biochemistry in mid-twentieth-century science.

In the early 1960s an eight-storey wing was added, more than doubling the size of the Department. The greater part of the necessary funds for the extension and its equipping came from the University Grants Committee, but I was told by the Registrar that if I required major research facilities I would have to find the money myself. I approached the Wellcome Trust, who gave the Department a grant of £50 000, making it possible to add an extra research floor of about five thousand square feet for the Head of Department. At that time, the cost of space was about £10 per square foot; now, in 1980, the cost per square foot is about £50–£60.

My research activities and those of my Medical Research Council team suffered no significant interruption by our move from Sheffield (as I have mentioned in Chapter 12). The great majority of people in the Department, including all its younger members, welcomed these skilful, experienced, and cooperative investigators from whom they could learn a great deal and receive much direct help. Later they were to tell me, 'The best thing you have done for the Department was to introduce this group of splendid scientists.' Luckily the objections from one or two of the Department's established members were not of great importance. It was perhaps natural for them to feel that my long and close association with the Unit carried a risk that I might neglect the other sections. I therefore went out of my way to care equally for all aspects of the Department's work. In particular I looked for outside support which, thanks to the generosity of the Rockefeller Foundation, and later the United States National Institutes of Health, made it possible for me to provide adequate facilities for everyone in the Department.

Soon after my arrival I was disturbed to find that in the science departments there were two 'classes' of lecturer,[25] the privileged and the underprivileged, differing in income and status. People in the privileged class had, in addition to their 'full-time' University Lectureship, a well-paid tutorial College Fellowship, which in extreme cases, almost doubled the holder's income. They also enjoyed a higher status and many amenities.[26] True, their Fellowships brought extra major responsibilities in the shape of tutoring and duties connected with admissions, examinations, and administration—all of which had to be discharged at the expense of research activities required by Statute as a Departmental duty. Research is very time-consuming, not only at the bench but also in necessary periods of quiet reflection, and because college duties make heavy inroads into Fellows' time, research suffers. As I have already mentioned, during the last twenty-five years very few Lecturer-Fellows in the experimental sciences—despite undoubted ability—reached a standard in research which would fit them for election to the Fellowship of the Royal Society.

The second group—the 'underprivileged class'—were those University Lecturers who had no major remunerative college attachment. However meritorious their research, however well they taught, however long the hours they worked, they received no recognition in terms of income or status. Often these 'mere' Lecturers played a specially important part in teaching, particularly at the post-graduate level, but post-graduate teaching received only nominal payment.

Other difficulties peculiar to Oxford were the rigid age-tie of the salary scale and the absence of the grade of Senior Lecturer. Thus it was impossible to recognize special merit by more rapid promotion or by promotion to a higher grade. Attracting good senior people to Oxford was difficult.

These problems were the outcome of some aspects of the autonomy of Oxford University. The total monies to cover salaries of lecturers and senior lecturers at any universities, and the general range of salary scales, are laid down by the University Grants Committee (the Government-financed agency which controls the flow of taxpayers' money to the universities). How this sum is divided is, however, up to an individual university. At Oxford the decision is made by Congregation, a body of all resident Masters of Art (numbering over 2200 in 1980). Voting by such a large body is liable to go against the recognition of merit for the simple reason that the majority cannot aspire to belong to the élite and thus they benefit from an even distribution of the funds available. The existence of a Senior Lecturer grade and accelerated promotion of the better people could only be achieved by cutting a sizeable slice from the salaries of the average people. So Oxford has no Senior Lecturers and an increase in pay comes, merited or unmerited, only with an increase in age.

All this became clear to me soon after my arrival because of the difficulties I met in attracting and retaining the best people, but I saw no opportunity to do anything about it until I had settled in Oxford and made myself fully familiar with its complex organization. In this connection it was of great help to me that I became a member of the Hebdomadal Council in March 1957.

I first raised the matter of abolishing the age-scale for Lecturers in October 1959, with the support of the Faculty Board of Biological Sciences, but the General Board decided not to take any action. The salary-income structure had many other critics besides myself. Our main worry was about its effect on the recruitment and retaining of good staff. These were the years of the 'Brain Drain' when good people were difficult to attract and only too easy to lose.

Hopes of improving the structure were raised in 1960 when, on 3 May, the Chancellor of the Exchequer announced in the House of Commons that because of the number of scientists leaving the country, he would make funds available through the University Grants Committee to universities which were having particular difficulty in attracting or retaining academic staff of adequate calibre. Two days later the University Grants Committee made an offer of £14 500 per annum to Oxford, 'to recoup . . . expenditure incurred in paying salaries above the normal in order to recruit or retain staff'.

While the University authorities took their time over considering the offer, the committee of Heads of Science Departments[27] on 21 October submitted to the General Board detailed information on the difficulties they were experiencing in staffing the Departments of Biochemistry, Botany, Engineering, Nuclear Physics, and Pathology. The committee stated, 'The evidence clearly indicates that action ought to be taken to relieve these difficulties and that the General Board and Council should be urged to accept the offer from the University Grants Committee.' The University authorities paid no serious attention to our submission.

Council referred consideration of the offer to the General Board. There were long delays in formulating a proposal to put to Congregation because opinions were divided and because there was a lack of sense of urgency. Eventually the Board decided, against the advice of its Finance Committee (I was told by a member) and by a vote of eight to six with eight absent or abstaining, to recommend rejection of the offer of money.

I felt that this recommendation should be openly challenged and wrote an article for the *Oxford Magazine*[28] (published 1 June 1961) from which I extract the following passages:

The General Board had before it reports on staffing difficulties in science departments. Some 40 scientists have in recent years refused to be considered for appointment, mainly because of the poor salary scales. Many candidates, especially those with family obligations, have stated in no uncertain terms that this is the main factor which keeps them away from Oxford. In my own Department several vacancies cannot be filled and some undergraduate teaching has to be done by non-University staff [i.e. members of the Medical Research Council Unit].

The offer of money is therefore most opportune. Why should it be declined when all other universities have thought it wise to accept? The General Board's reason, as printed in the *Gazette*, is the following, 'The General Board did not see how the criterion of market value could be introduced at this stage without creating unacceptable inequalities of treatment.' At first this may appear a valid point but I submit that it does not stand close examination. Scarcity and merit go together. The University would be in a position to award money according to merit. If we had a truly egalitarian society within the University, where all Lecturers of the same age received the same income, one might well object. But as long as the enormous inequalities arising from College emoluments exist, any argument against salary flexibility on the grounds of equality is fatuous.

My argument was strengthened by an article in the following week's edition of the *Magazine* (8 June) in which an anonymous lecturer called attention to the way that recruits to Oxford from outside (like himself) with no effective college affiliation, were treated as 'useful second-class citizens'. For such Lecturers, life at Oxford offered 'continual frustrations, financial, social and others connected with teaching'. At first, said the writer, it had appeared to him to be the 'reflection of a fine, noble and egalitarian society that all Lecturers of the same age are paid at the same rate. It therefore comes as a surprise that another Lecturer receives several hundred pounds a year extra because he holds a College Fellowship . . . and substantial financial benefits in addition to his Fellowship stipend.' Commenting on this letter, the *Observer* (18 June 1961) said, 'This complaint about Oxford life is not new, of course, for Sir Roy Harrod in his life of Lord Cherwell (1), writing of conditions a generation ago, remarked, "Even a well-paid laboratory appointment did not raise you above the helot class unless a Fellowship was attached".'

On 13 June Congregation voted. The result was, alas, predictable and the offer of the University Grants Committee was turned down by 163 votes to 106; only one member in five voted.

I found the weak arguments of the General Board and their acceptance by Congregation deeply disturbing, a blow to those who aimed at attaining and maintaining high academic standards. Congregation's rejection I saw as being due to a combination of various attitudes. The majority of members were indifferent to the outcome and had not turned up; there was among some members, I believe, the attitude of 'I'm all right Jack'; but principally I believe there was a general failure to recognize the importance of science to the national wellbeing and economy. Perhaps there was still something left of the attitude which Roy Harrod (1) referred to as being prevalent in the 1920s. 'Cherwell,' wrote Harrod, 'felt that among the humanists of Oxford there was great complacency and a profound ignorance about the intellectual status of science. He liked to tell the story of one of his first dinner parties in Oxford. He had been bidden, as a new Professor, by the Warden of All Souls, and, finding himself sitting next to the Warden's wife, expressed his misgivings about the status of science in Oxford. "You need not worry", she assured him, "A man who has a first in Greats[29] could get up science in a fortnight". When he told this story in later years, he credited himself with replying, "What a pity that your husband has never had a fortnight to spare".'

Though this round of the battle had been lost, I felt loath to give up the fight, for two main reasons. First because the University salary scale—in some respects the lowest of any British university—was failing to attract enough first-rate scientists; secondly, because the offer of money was refused on the grounds that it would introduce inequality and to me the then existing—and still present—income structure (note the difference between salary and income) was characterized by gross inequality. So I wrote a second article for the *Oxford Magazine*, and this was published on 19 October 1961. In it I detailed the inequalities between the incomes of non-Fellows and Fellows, who received extra income from tutoring plus allowances towards housing, entertainment, book buying, and other smaller expenses. I pleaded especially the cause of science in that good scientists were being seduced from their benches to accept lucrative College Fellowships. The idea that the duties of a College Fellow could somehow be discharged by full-time University Lecturers in 'spare time' I found quite unacceptable.

My arguments were fiercely attacked in the *Magazine* on 26 October by Dr G. A. F. Chilver, a fellow Member of Council, mainly on the grounds that there was not sufficient information to support my claims. Chilver took me to task on several specific points and challenged my assertion that it was 'monstrous to suggest that a Lecturer involved in research using machines has any spare time', even though Chilver, as a Council Member, had before him a memorandum from Professor Wilkinson (Nuclear Physics) stating:

If a man is prepared to immerse himself deeply in College teaching it effectively neutralises him so far as research is concerned, at least during term time. Research in nuclear physics, which is carried out almost entirely on accelerating machines, with smaller or larger groups of workers in collaboration, must be organised and scheduled several weeks or even months in advance. It is impracticable to suggest that the research can be arranged to fit in, day by day, with College teaching. A large accelerator costs more than £1000 a day to run and is the object of the fiercest competition between teams. The suggestion that research on these machines has to be organised for the convenience of undergraduate teaching would be laughable and result, very simply, in the research team in question getting no time on the machine. It is therefore utterly impracticable to suggest that nuclear physics research on such machines can be done in odd hours or odd days to suit tutoring.

My article did not give rise to any action, but the subject matter was taken up in 1963 in the Report of the Committee on Higher Education (Cmnd. 2154), set up by the Prime Minister and Chaired by Lord Robbins. Paragraph 687 of the Report referred to Oxford and Cambridge and included the following comments:

The general obscurity in which so many of their administrative and financial arrangements are shrouded is not compatible with the situation in which they, like other universities, are largely dependent on public funds. Continuance of such anomalies may well endanger not only their own welfare, but also the effectiveness of the whole sphere of higher education in this country of which they are and should be so splendid a part. We are aware that in both universities these problems are being considered and solutions sought. We recommend that, if Oxford and Cambridge are unable satisfactorily to resolve their problems within a reasonable time, they should be subject to an independent enquiry.

In response to these criticisms Oxford, rather than face an independent enquiry from outside, set up a 'Commission of Inquiry' under the Chairmanship of Lord Franks. The 'Franks Commission' invited comments from members of the University as well as from outsiders. It received one million words of written comment and one and a half million words in oral evidence. In 1966 the Commission issued a report (2) of their findings, their critical analysis and their recommendations. These comments, some highly critical, were published, but not in a form readily accessible outside the University. Only a few of its 170 useful recommendations have so far been implemented. Two new Colleges, St Cross and Wolfson, were established and these provided Fellowships—non-stipendiary Fellowships. They gave people status and a few minor privileges but they did not abolish the large discrepancies in incomes. One of the Commission's recommendations (Paragraph 400) which was not accepted by the University was that, in order to come into line with the practice of other universities, post-graduate supervision should be regarded as part of the normal teaching duties of full-time staff, i.e. not as an extra fee-earning activity. This would have saved the University some £36 000 in 1964–5.

The conditions for many Lecturers have not changed fundamentally to this day. As a result of much outside pressure and inside unrest the number of non-Fellows is less, but gross differences still exist. There are still many members of the teaching staff who feel unfairly treated. Their dissatisfaction is however seldom aired because, I believe, they feel that life could be made very awkward if they were to raise matters unpopular with the financially privileged. Not infrequently have I heard the comment, 'Oxford is not the place for me' from young people who resent the class distinctions within the University.

There are many other valuable recommendations in the Franks Report[30] which have not even been discussed. But after all, fifteen years is very little in the life of an institution that is over seven hundred years old. I know I have a tendency of sometimes being unduly impatient and too eager to get on with the job (an attitude which sometimes at least I have found useful); the attitude of many other senior people in the University was expressed by the Editor of the *Oxford Magazine* on 17 February 1967. In response to my suggestion that Paragraph 400 of the Franks Report should be implemented forthwith, he wrote, 'This recommendation has been much discussed among Dons but it is not yet clear that any decisive opinion has been formed on the matter. We would estimate that whatever the final decision on this might be, those who support the Recommendation would not be furthering their case in the long run by rushing in too quickly. Council has before it a very long programme of legislation which will keep it occupied into the autumn of this year. We are not just pleading the convenient doctrine of "unripe time"[31] when we suggest that this complex question should be dealt with at a later stage.' The Board was acutely aware of the problem because many members of the Faculty were in the underprivileged class, but the General Board decided to take no action; its members all belonged to the privileged class.

Another feature of Oxford which I thought required some changes was the pyramidal structure of the large science Departments, wide at the base with one professor at the top. A pyramidal structure may still be appropriate for small departments but in the late 1950s the Oxford Biochemistry Department was already relatively large and growing quickly. At Harvard in 1951 I had seen 'multi-professor' departments with a number of relatively small units, each headed by a full professor, located in the same building. These groups were financially largely independent but collaborated closely in teaching and administration. Responsibility for general administration was taken by a chairman—an office held for a period by each professor in rotation.

This kind of organization was very much in keeping with the changing spirit of the times and would afford Oxford two main advantages. It would make more room at the top—a very important matter for the morale of younger academics who saw too little scope for advancement. Secondly, for the most effective running of a laboratory it is of great importance that top people are on the spot, and the 'full-time' Lecturer-Fellows, being too busy with college duties, are liable to be conspicuous by their absence. Professors at Oxford, on the other hand, have no major time-consuming college responsibilities, and can devote virtually all their time to departmental work, strengthening its activities in teaching and research.

A chance to create a second Professorship in my Department arose within a few weeks of my arrival. There was within the Department a highly qualified biochemist, Donald Woods, who held a Readership. Woods had made outstanding contributions to microbiology, among them the discovery, in 1940, of the mechanism of action of sulphonamides—a landmark in the development of both biochemistry and chemotherapy. The therapeutic efficiency of sulphonamides had been found empirically and until his discovery it was not known why they were effective. Woods showed that the sulphonamides (which have the general structure (H_2N- ◯ $-SO_2NHR$) interfere with the metabolism of an important cell constituent, p-aminobenzoic acid (H_2N- ◯ $-COOH$)) by virtue of their similarity in chemical structure. The concept of 'competitive' interference has since been a guiding principle in attempts to develop a planned chemotherapy and is one of the few basic principles on which rational chemotherapy rests.

In November 1954 I learned that other universities (Nottingham and London) were trying to attract Woods by offering him the Headship of a Department of Biochemistry. I and many others were very anxious to keep him at Oxford, and an opportunity of making Oxford more attractive to him arose from correspondence which he had with Mr N. B. Smiley, the Managing Director of Arthur Guinness, Son & Company (Park Royal) Limited. Because of Woods's achievements and promise, this company had supported chemical microbiology at Oxford, under Woods's direction, since 1952 when they had established three Guinness Research Fellowships and a personal Guinness Fellowship at Trinity College for Woods. The Research Fellowships were post-doctoral and intended to attract mature researchers to Woods's laboratory, but it proved difficult to attract the right kind of worker at the salary offered. Woods had therefore suggested to the company that the number of Fellowships might be reduced to two, so that the stipend could be increased. Mr Smiley came to Oxford on 6 May 1955 to discuss this with Woods and myself. Guinness, he said, did not wish to reduce the number of Fellowships. On the contrary they were very willing to increase their support for Woods in view of his high standing. I suggested that his Readership should be raised to a Professorship, and that this might be named after the Chairman of Guinness, Lord Iveagh. Mr Smiley felt that this would be most agreeable to his company.

By 26 May, discussions with the company had reached the stage where I was able to put a proposal to the Board of Biological Sciences for Woods's promotion to Professor of Microbiology on the understanding that Guinness would meet the extra cost. In accordance with precedent, the Professorship would be of the so-called 'Schedule A' category (Schedules A and B differ in respect to salary and administrative responsibility). The Faculty Board accepted the proposal and submitted it to the General Board who, alas, delayed a decision for five months, during which time Woods received repeated encouragement to go to London University. The Board was 'disturbed'—a lame reason for the delay—that my proposal might have repercussions on the Schedule B Professors of Entomology, Byzantine and Modern Greek, Egyptology and Eastern Religions, and Ethics. Not until November, after seemingly endless arguments which later turned out to have been futile, was Woods appointed 'Iveagh Professor of Chemical Microbiology', with the stipend and duties of a Schedule A Professor—exactly as I had proposed in May. It was a source of great satisfaction, not only to everyone in the Department of Biochemistry, but to innumerable scientists elsewhere. Justice had been done to a man of exceptional distinction.

The next opportunity to further my hopes for multiprofessor Departments at Oxford came in November 1960 when the University Grants Committee visited Oxford for discussions on the next quinquennial grant, and they asked to talk with Heads of the Science Departments. The majority of the Heads were anxious to see new professorships established in the larger Departments and took the opportunity to discuss their ideas with the University Grants Committee. The Committee listened sympathetically and the Heads of Science Departments prepared a memorandum for submission to the Hebdomadal Council, outlining our proposals and giving our reasons:

The Heads of Science Departments have two considerations principally in mind. First, the amount of departmental work falling to the professor has increased greatly, and in some departments at least can no longer be handled without loss of efficiency. This work arises from the increased size of departments, the increasing complexity of administration, the increasing need to collect funds from outside sources, and the increasing call for service outside the University, on Government committees, Research Councils, Foundations, learned societies, and examining at other universities. The load will continue to grow unless the number at Oxford of those of professional status increases at the same rate as scientific activity generally.

Second, there is a need for more professorial appointments to allow new fields to develop, and to provide room for younger people of high calibre whose ability to lead a group may only find an outlet at present by migration to the USA.

The Heads also pointed out that the ratio of students to professors at other British universities ranged between 43 and 73 students to one professor, while the ratio at Oxford and Cambridge was 99–100 to one.

The memorandum continued,

The Heads of Departments considered whether the needs could be met by appointing more Readers. Two reasons prompted them to conclude that this idea was less advantageous. First, Readers may have considerable college duties and many have demonstrating duties, while some have both. They may therefore not have the time to take over major responsibilities of departmental administration. Second, hitherto, in the eyes of the University Grants Committee the number of readerships at Oxford is already high in proportion to less senior posts; this cannot be said of professorships.

Council referred the paper to the General Board, which flatly rejected it. Council did not accept the General Board's reasons and set up a joint committee with members from both bodies. After a hard and lengthy fight, this committee agreed on proposals for new professorships. The University Annual Report for 1962–3 took 'pride in the progressive development which the creation of new professorships envisages'.

But when it came to carrying out the new development, the General Board very badly mismanaged and, I fear, destroyed it. It did not act in the spirit of the original proposals but instead regraded a number of Readerships to Schedule A Professorships, including eventually some members of its own ranks! But there were no new posts. The original intention of creating additional posts to provide promising younger men with opportunities and suitable status was ignored. No extra room was made at the top. No new field was opened. No new scientist was attracted to Oxford. No-one was dissuaded from emigrating. To give senior distinguished Readers the title and salary of Professor after long years of service may be a splendid gesture, and may be well deserved, but it did not, in 1961, meet our plea for more room at the top.

The Faculty Board drew attention to this miscarriage of the intention but the General Board refused to admit the validity of these comments—not unexpected when the accused is also the judge.

Another of my major administrative preoccupations at Oxford was the problem of introducing molecular biology, an important and fast-developing subject, into the University. This eventually resulted in the setting up of the Laboratory of Molecular Biophysics.

I first discussed it with Dorothy Hodgkin in 1958. We were convinced that the kind of work being done in the Molecular Biology Unit at Cambridge, led by Max Perutz, was of importance not only to research in biology but also to undergraduate and post-graduate teaching. It had come to our knowledge that the Cambridge team might, in the not-too-distant future, find itself without a home (Cambridge University had indicated—an extraordinarily short-sighted decision—that it could no longer give them space) and we had hopes of attracting the team or some of its members to Oxford. Dorothy and I discussed this possibility with Max Perutz and John Kendrew, but shortly afterwards the Medical Research Council decided for once to abandon their general principle not to invest in bricks and mortar and to build a laboratory for the Unit at Cambridge. The subsequent successes of the Laboratory, which included four Nobel Prizes (Perutz, Kendrew, Crick, and Sanger) bear witness to the wisdom of this decision.

An alternative idea to build up molecular biology at Oxford, raised by Dorothy also in 1958, was that we might, in time, attract David Phillips, an X-ray crystallographer whom she considered to be of high promise and who was then working on haemoglobin in Sir Lawrence Bragg's group at the Royal Institution. Phillips's long-term future was uncertain since no-one knew what would become of the group when Bragg, then sixty-nine, should decide to retire. Dorothy felt, however, that we should delay an approach to Phillips until we were in a position to offer him adequate facilities. Four years later, in 1962, Dorothy and I discussed with Phillips the possibility of his eventual transfer to Oxford with his team. I undertook to discuss matters with my senior colleagues in the Department of Biochemistry and possibly to explore the reaction of the University, and to find out from the Medical Research Council if they would be prepared to collaborate on a transfer.

The discussions with my colleagues revealed that they considered a further growth of the already large Biochemistry Department undesirable; they also had reservations about the usefulness of Phillips's team in undergraduate teaching.

Later in 1962 a new avenue opened up when the University agreed to a request by John Pringle, the new Professor of Zoology, to establish a Readership in Biophysics in the Zoology Department. No suitable candidate was willing to take this position on the terms Oxford offered, and Pringle said that he would try to get support from one of the Research Councils to set up a Biophysics Unit, with himself as its Director, which would have the scope and facilities to attract a good biophysicist as Reader.

I gave Pringle detailed particulars of David Phillips and of the discussions Dorothy Hodgkin and I had had with him. It seemed sensible to both Pringle and myself that a Readership in

Biophysics set up in the Department of Zoology could work in close collaboration with my own Department and become the nucleus from which an independent Department of Biophysics could eventually be established.

Pringle wrote to Sir Harold Himsworth, Secretary of the Medical Research Council, in February 1963 putting these ideas forward and asking whether Himsworth would consider the feasibility of the transfer of Bragg's team, or a large part of it, to Oxford at some future date. Before he received a reply, Pringle got in touch with Phillips and was very impressed by the man and his work.

At about the same time, Dorothy Hodgkin talked to Sir Lawrence Bragg who told her that he intended to retire in 1967. There was a general feeling, she said, shared by the Medical Research Council, that Bragg's team should not be broken up before he retired. She also suggested that immediate links might be established between Oxford and the Royal Institution group by inviting Phillips to give regular courses of lectures at Oxford, perhaps within the Honours School of Biochemistry.

I have no note in my files of how the next step came about, but they record that Bragg suggested that I should visit him and that I replied, fixing a day for the meeting and suggesting that I should take Pringle with me. Pringle and I wrote separately to Bragg in advance of this meeting, setting out our ideas for the organization of the future development of biophysics at Oxford.

When we met on 12 June 1963, Bragg, now seventy-four years of age, said that he would be delighted to 'marry off his children' when he retired in a few years' time, by making arrangements for them elsewhere, perhaps at Oxford.

By October, Sir Harold Himsworth was able to tell us that the Council had received our proposals favourably and was satisfied that Bragg was enthusiastic for his team to move to Oxford.

But Himsworth also told us that we were not the only people interested in attracting Phillips and his team: approaches had been made by other centres in Britain and overseas. Oxford must therefore, he said, make a formal proposal. He himself would prepare the ground to have preliminary talks with the Vice-Chancellor and Registrar during December or January.

Our hopes were to get approval from the University for a Professorship for Phillips since the Biology Board—as a matter of principle and without reference to names—considered biophysics as a first priority among various proposals which had been put before them. Pringle expected to be able to make adequate space available.

Up to this time, after five years of exploring the possibilities, everything had been done without the knowledge of formal University bodies. Now the time had come to get things moving on an official basis and to make arrangements for Phillips and his team to give lectures and demonstrations to undergraduates in the Honours Schools of the biological subjects. The preparation of Departmental proposals for the quinquennial report for 1967–72 gave us the opportunity to raise our plans formally.

Almost every Department had rival suggestions to make about new Chairs, but agreement was reached that Molecular Biophysics and Cellular Physiology should be given highest priority, and a memorandum to this effect approved by the Biology Board, was submitted to the General Board. In November, the Biology Board received a further memorandum from Professor Pringle, amplifying the earlier proposal and mentioning for the first time in an official university document the name of Dr D. C. Phillips. Pringle recommended that Phillips, along with other candidates, should be interviewed for the appointment, which should begin the following October.

For almost eleven months, the Biology Board was unable to further matters. Not until 26 October 1964 did it receive a communication from the General Board about the proposed professorship, with an appended memorandum from Phillips describing the financial implications of the move. The General Board ruled that the proposed professorship could be regarded as the upgrading of the existing post of Reader. It was therefore prepared to finance a professorship, from October 1967 (three years ahead!) on the clear understanding that 'provision could be obtained from outside sources for the "small number of more junior posts" and "facilities on an adequate scale"'.

A report on the detailed requirements of finance and accommodation was prepared by the Buildings and Developments Committee. Having considered the report, the General Board felt that it would be desirable to obtain the views of the Faculty

Board about the scale of the proposals in relation to other developments. The result of the Biology Board's deliberations was that, welcome as the development of Molecular Biophysics would be, it might seriously prejudice other urgent claims, such as the establishment of the Chair of Cellular Physiology and further University Lecturerships. Before final decisions were made, the Board suggested, further negotiations with the Medical Research Council should take place to discuss the times and terms of Oxford's takeover of Phillips and his colleagues, and so the Vice-Chancellor wrote to Sir Harold Himsworth on 2 November. Parallel negotiations of a preliminary nature had been under way between Professor Pringle and the Nuffield Foundation for support to house Phillips's project in the new Zoology building.

Sir Harold Himsworth replied to the Vice-Chancellor two days later. His opinion was that although the Council were, in principle, sympathetic, they would need some time to assess what support they should provide for such a major undertaking. Himsworth added that he would set detailed investigations in hand immediately, to be ready for examination by the Council early in the New Year. He also declared his willingness to answer any questions which the Nuffield Foundation wished to raise.

Although disappointed by these further delays, Pringle and I contained our impatience and made no further approaches or enquiries to Himsworth, being reasonably confident that our proposal would have a fairly easy passage at the Council meeting.

The Biology Board had Himsworth's answer on 2 March. The Council was willing (subject to the agreement of Oxford University to apply to the University Grants Committee to take over the project 'at a time to be agreed') to provide some £30 000 a year for research (but not for teaching) plus substantial contributions for equipment. They were also prepared to advise the Nuffield Foundation that the proposed accommodation should be built even if it could not all be put to immediate use.

The remaining stumbling block (readers will not be surprised to learn) was that of being able to offer two key members of Phillips's team an acceptable salary. If they made 'level transfers' to the age-tied Oxford scales, they stood to lose several hundred pounds a year. The old story. Extra emoluments through College Fellowships provided the only hope of making up the difference. The Medical Research Council had no power to endow Colleges, nor did any College offer Fellowships for these men from their own resources. It fell to Sir Lawrence Bragg to go begging for funds which could be channelled to these two men through a College Fellowship. Lincoln and Wolfson accepted the monies raised by Bragg and appointed the two as Research Fellows.

I should perhaps point out that the majority of the people making these decisions for the University had no direct knowledge of some of the issues at stake. In particular, few were in a position to assess the unique significance of Phillips's work or how important it was that his team should stay together. The work which led to the elucidation of the structure of lysozomes was not finished until after the main negotiations had been completed, and its publication, in *Nature*, did not take place until 22 May 1965. It was at once hailed internationally as a great pioneering achievement—the first successful detailed analysis of the three-dimensional structure of an enzyme. The impact on the world of science was tremendous and Oxford was justly proud of being associated with the discoverers.

The difficulties which had to be overcome to establish this new Department at Oxford are characteristic of most major innovations in the sciences. Similar stories could be told about other Departments: for instance, those recently established in the Faculty of Medicine, i.e. Cardiovascular Medicine, Clinical Neurology, Paediatrics, Chemical Pathology, Clinical Pharmacology, Psychiatry, Radiology, and Social and Community Medicine. In all these cases the University accepted the proposals on condition that a large endowment was made.[32] In 1980 the sum requested for the endowment of a new University professorship in the sciences is £400 000.

New moves are rarely initiated by the official bodies governing the University. They usually come about because one or two individuals prepare a case in great detail before they put it to a Board. However worthy, new proposals have little chance of success unless they can be shown to be financially possible, i.e. that the proposers have reasonable prospects of outside support, otherwise they would be turned down on the grounds that there were numerous equally deserving cases competing for the University's limited funds. A brilliant idea, brilliantly expounded, is not enough to bring a major new development to fruition. At Oxford brilliant ideas are common, as common as pebbles on the beach—but there are also many brilliant people to demolish any brilliant idea. Ideas, unless backed by cash, are liable to evaporate into nothingness.

These are some of the experiences I had when trying to maintain academic standards. They illustrate the difficulties that arise from over-democracy, over-bureaucratic self-government and the absence of individuals with the overall personal responsibilities of the kind a Prime Minister and Cabinet Ministers have in a political democracy or the Managing Director and Board of Directors have in business organizations. Additional handicaps are imposed by the personal bias, even self-interest, of those participating in University government. In political democracy as practised in Britain, all members of committees who are 'interested parties' must declare their interest and abstain from voting, and the rules on what constitutes 'personal interest' are very strict. We have it from the authority of an Oxford Vice-Chancellor of the 1960s, Sir Kenneth Wheare (3), that 'interested parties' covers all those who are closely identified with an organization or a point of view and who, in this capacity, take part in British administration. This category, Wheare states, may include, say, 'a woman who is secretary to the Parish branch of a pacifist society and is called upon to give evidence before a Royal Commission on the arms traffic'. It is true, the Hebdomadal Council and the General Board have a Standing Order under which a member is required to declare a personal interest, but this does not cover all relevant interests, and there are many committees in the University which either have no such Standing Order or do not adhere to it rigidly enough. Altogether the practices of political democracies are not, and cannot, be applied to democracy within a university. Political democracy in Britain means government by elected representatives; these representatives elect a leader (the Prime Minister) who appoints the Cabinet, which is given great powers. But these powers are tempered with accountability and the possibility of dismissal by Parliament or by the popular vote.

There is no equivalent to these arrangements in the University, although the Hebdomadal Council is sometimes referred to as 'the Cabinet'. The Hebdomadal Council enacts legislation which is laid before the whole Congregation for approval or rejection, but only on very special occasions does Congregation discuss legislation. A major difference between political democracy and University democracy is the absence in the University of a post comparable to that of a responsible, well-informed Minister specializing in specific areas and supported by a first-rate team of Civil Servants. Members of the Hebdomadal Council and the General Board are all very able people but many are shockingly uninformed about subjects on which they are asked to form an opinion and to vote. In earlier days, when the University was a relatively simple organization and entirely self-financed, the competence of the Hebdomadal Council was adequate. I doubt whether it is still so today in the area of science, pure and applied. It is no accident that three times in its recent history (1850, 1872, and 1922) Royal Commissions had to be appointed to set Oxford's house in reasonable order. The possibility of another Royal Commission was raised by the Robbins Committee in the 1960s, but the University managed to avoid this intrusion by appointing the Franks Commission. The governing bodies as they are now constituted have neither sufficient foresight, imagination, initiative, nor motivation to ensure the University's continued rejuvenation. Indeed these so-called 'democratic' bodies are liable to be highly conservative. Consciously and subconsciously, self-perpetuation is the tendency. In 1908 Francis Cornford described this attitude in his oft reprinted *Microcosmographia Academica* (4), a wise and most amusing 'Guide for the Young Academic Politician'. Cornford formulated several great principles that prevent progress in the academic world:

the Principle of the Wedge: You should not act justly now for fear of raising expectations that you may act still more justly in the future—expectations which you are afraid you will not have the courage to satisfy.
the Principle of the Dangerous Precedent: Nothing should ever be done for the first time;
the Principle of the Unripe Time: People should not do at the present moment what they think right at that moment, because the moment at which they think it right has not yet arrived.

There is, says Cornford, only one argument for doing something, namely that it is the right thing to do—but then the results of any course of action are so difficult to foresee with certainty, or even probability, that the only justifiable attitude of mind is suspense of judgment.

A major obstacle at Oxford to frank criticism of matters involving people, especially if it touches their self-interest, is the risk one takes that life could be made rather difficult if criticisms have to be made in public. The Franks Commission refused (without giving any convincing reason) to admit evidence submitted confidentially, in other words it chose to arm itself with blinkers.

My criticisms of Oxford University may prompt the question, 'How then has Oxford maintained its high standards?' My answer would be that the machinery for filling senior University posts (Professors and certain types of Reader) has been very effective. The Electoral Boards are small (until recently they had nine members, now a few more) and several members are experts, not only highly competent but willing to take a great deal of trouble to find the right candidate. Moreover, self-interest does not enter. Occasional mistakes are, of course, inevitable, but these do not affect the overall standard.

These Boards alone make the decisions. There is no complicated bureaucracy, no involvement of Faculty Boards, General Board or Hebdomadal Council. Similarly the election of College Fellows is unbureaucratic and in competent hands—those of the Governing Body, sometimes advised by a subcommittee. Great care is taken in choosing the right people, with special reference to their competence in and devotion to undergraduate teaching. Since the appointment of the right people is by far the most important single part of University administration and the most important factor in maintaining standards, the present election procedures have been a saving grace.

These administrative matters did not interfere seriously with my teaching and research work. My Medical Research Council team accepted many D.Phil. students and visitors from overseas and, in the thirteen years from 1954 to 1967, published many papers in the following fields.

In 1957, Hans Kornberg, with my collaboration [167], discovered a modification of the citric acid cycle in which the stages between isocitrate and succinate are replaced by two other reactions. Isocitric acid, instead of reacting to form α–oxoglutaric acid, as shown in Chapter 8 page 115, is split into succinic acid and glyoxylic acid:

$$\begin{array}{c}\text{COOH}\\|\text{H}\\\text{C}\\|\diagdown\text{CH}_2.\text{COOH}\\\text{CHOH}\\|\\\text{COOH}\end{array}\quad\longrightarrow\quad\begin{array}{c}\text{COOH}\\|\\\text{CH}_2\\\diagdown\text{CH}_2.\text{COOH}\end{array}\;\text{Succinic acid}$$

$$+$$

$$\begin{array}{c}\text{HCO}\\|\\\text{COOH}\end{array}\;\text{Glyoxylic acid}$$

Isocitric acid

This is followed by a reaction of the glyoxylic acid with acetyl coenzyme A:

Glyoxylic acid
$$\begin{array}{c}\text{COOH}\\|\\\text{HCO}\end{array}$$

$$+$$

Acetyl Coenzyme A
$$\begin{array}{c}\text{CH}_3\\|\\\text{CO}\\|\\\text{Coenzyme A}\end{array}$$

$$\longrightarrow\quad\begin{array}{c}\text{COOH}\\|\\\text{CHOH}\\|\\\text{CH}_2\\|\\\text{COOH}\end{array}\;\text{Malic acid}$$

$$+$$

Coenzyme A

When these two reactions replace the stages of the citric acid cycle between isocitrate and malate (compare Fig. 2, p. 115) the result is

$$\begin{array}{c}\text{COO}^-\\|\\\text{CH}_2.\text{CH(OH)}.\text{COO}^-\\|\\\text{CH}_2\\|\\\text{COO}^-\end{array}\quad\longrightarrow\quad\begin{array}{c}\text{COO}^-\\|\\\text{CH}_2\\|\\\text{CH}_2\\|\\\text{COO}^-\end{array}\;+\;\begin{array}{c}\text{O}\\\|\\\text{HC}.\text{COOH}\end{array}$$

Isocitrate   Succinate   Glyoxylate

$$\text{HC.COO}^- + \text{CH}_3.\text{CO.SR} \rightarrow \begin{array}{l} \text{COO}^- \\ | \\ \text{CHOH} \\ | \\ \text{CH}_2 \\ | \\ \text{COOH} \end{array} + \text{HSR}$$

Glyoxylate   Acetyl CoA   Malate   CoA

The sum of these two reactions is

isocitrate + acetyl CoA → succinate + malate.

In Fig. 9, substances entering the cycle are written within the circle, those leaving are written outside.

The overall effect of the modified citric acid cycle is fundamentally different: two molecules of acetyl CoA plus half a molecule of O_2 plus two molecules of H_2O form one molecule succinate plus two molecules coenzyme A

2 acetyl CoA + ½ O_2 + 2 H_2O → succinate + 2 coenzyme A

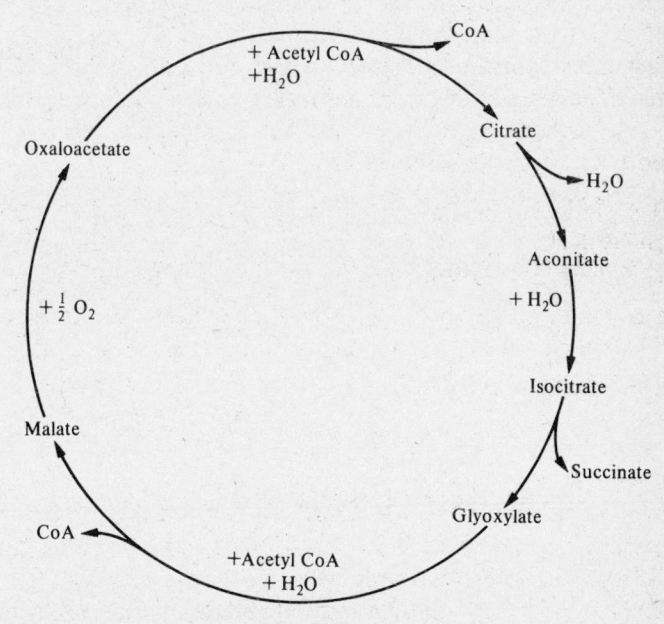

Fig. 9. Diagram of the 'glyoxylate cycle'.

This new cycle occurs in micro-organisms and in plants.[33] Smith and Gunsalus had already described the presence of 'isocitrase' in the bacterium Pseudomonas in 1954 (5); Wong and Ajl discovered 'malate synthase' in extracts of *Escherichia coli* in 1956 (6). Neither team, however, had been able to define a physiological significance for the two reactions. Kornberg and I asked ourselves how certain micro-organisms could build up all cell constituents if acetate were the sole organic substance in the culture medium. Numerous experiments with carbon-14-labelled acetate revealed that within seconds the intermediates of the citric acid cycle were detectable but that no other substances became labelled. The glyoxylate cycle can account for this.

Pathways from succinate and acetyl CoA to the synthesis of many cell constituents were already well established and the glyoxylate cycle could thus be a major starting-point for the synthesis of cell constituents. In many plants, especially in seedlings from oil-containing seeds, it is a mechanism of the transformation of fat into carbohydrate. Higher animals cannot form carbohydrate from fat. Plants (the first experiments were carried out on jack beans and pumpkin seeds) break down storage fat to form acetyl CoA, which through the glyoxylate cycle is converted via succinate to carbohydrate, such as cellulose, and other cell constituents. The stages between fat and carbohydrate are as follows:

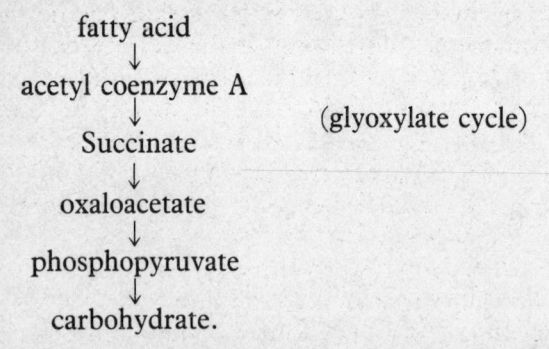

$$\text{fatty acid} \\ \downarrow \\ \text{acetyl coenzyme A} \\ \downarrow \\ \text{Succinate} \quad \text{(glyoxylate cycle)} \\ \downarrow \\ \text{oxaloacetate} \\ \downarrow \\ \text{phosphopyruvate} \\ \downarrow \\ \text{carbohydrate.}$$

During his stay with the team, Rodney Quayle started his work on the chemical processes by which certain micro-organisms can derive all their carbon requirements from simple one-carbon compounds like methane (CH_4), methanol

(CH₃OH), formaldehyde (HCOH), and formic acid (HCOOH). The micro-organisms synthesize, from these substances plus ammonia and salt, all cell constituents necessary for life. This work proved to be not merely an academic exercise but to have industrial applications (7, 8, 9). In the late 1960s Rod participated in a symposium on hydrocarbon microbiology organized by the British Institute of Petroleum, reporting his work on the synthesis of cell constituents for methane. The audience included two representatives of Imperial Chemical Industries (ICI) and one of them, Peter King, spoke to Rod after his lecture, suggesting that there might be technological applications of his work since methane was now abundantly available from the North Sea gas fields. He invited Rod to visit the ICI works at Billingham to discuss the matter.

In the event, Rod became a consultant to ICI and Peter King developed, with great energy and resourcefulness, an industrial process to produce protein for animal feeding stuffs from methane, ammonia, and a few salts (10). In the late 1970s ICI invested forty million pounds in a plant for commercial production. The story is a splendid example of the practical relevance of basic research.

Kenneth Burton made important contributions to the biochemistry of nucleic acids. One, the development of a new method for the estimation of DNA, was adopted by hundreds of other investigators (11). With Benzinger, Kitzinger, and Hems, Burton established the value of one of the most important thermodynamic constants in biology, the free energy of hydrolysis of MgATP, a value needed for many calculations (12). Under standard conditions at pH 7.0 and 37°, the value was found to be -7.0 Kcal. At the concentration prevailing in living cells it is usually between -10 and -12 Kcal. The value was obtained by ingenious measurement of the equilibrium constant of the glutamine synthetase reaction.

Other research activities can only be mentioned briefly. Walter Bartley made important contributions to mitochondrial metabolism, and Ronald Whittam to ion transport. David Hughes developed into a highly original investigator in the field of pure and applied microbiology. My own group (Leonard Eggleston, Reginald Hems, Derek Williamson, and Patricia Lund) turned its attention from metabolic pathways to regulatory mechanisms of metabolism. We studied gluconeogenesis (the synthesis of carbohydrate from noncarbohydrate precursors). By developing new enzymic methods [187] for the quantitative determination of ketone bodies (since universally adopted in this field), we studied their physiological role in the animal body and established the concept that ketone bodies serve, as do glucose and fatty acids, as energy sources for several tissues, especially in the heart when the diet is low in carbohydrate or during fasting. We developed methods for the study of metabolism in the isolated perfused organ (liver, [216, 223] kidney, [226] muscle (13)) of small laboratory animals, mainly the rat. Earlier perfusion of isolated organs could only be carried out on larger animals, such as cats and dogs, for two reasons: the sample of perfusion fluid obtainable from smaller animals was insufficient for many types of quantitative analyses of metabolites and there were difficulties in cannulating the narrow blood vessels of small animals. By the later 1960s the perfused organ had largely replaced tissue slices and minced or homogenized tissues. To study the subtle regulation of metabolism, it is important that the cells are in their natural environment. During the 1970s the use of perfused organs was partly replaced by that of isolated liver cells and isolated fragments of renal tubules; one of the main initiators of the successful preparation of isolated hepatocytes was Michael Berry (14), who had earlier been a member of my team, but the decisive experiments came after he had left Oxford.

Another area of extensive investigation initiated at Oxford by Derek Williamson, following a lead given by Bücher and Klingenberg, was the measurement of the redox state of the pyridine nucleotides, i.e. the relative concentrations of free NAD and $NADH_2$. The ratio of the concentrations of these enzymes is a key factor in the control of many metabolic processes [225].

Those interested in details of our work will find references in the list of publications.

I must emphasize that this survey covers only the work of my Medical Research Council Unit and those closely associated with it. Much outstanding research work was carried out independently by other members of the Department of Biochemistry: Donald Woods, Sandy Ogston, R. B. (David) Fisher, Denis Parsons, Paul Kent, Lloyd Stocken, Margery Ord,

Rupert Cecil, June Lascelles, R. M. (Morrin) Acheson, Arthur Peacock, Ian Walker, Charles Pasternak, Keith Dalziel, Mary Lunt, and Michael Foster, who between them also carried the chief weight of undergraduate teaching.

18
'RETIREMENT'

My appointment with Oxford University had to terminate on 30 September 1967. Oxford University (like other British universities) does not support staff beyond retirement age. This is a matter of principle and is related to the democratic control of University appointments by committees consisting entirely of University staff. No academic appointments can be made at Oxford by executive officers as they are by the Deans or Presidents of American universities. At Oxford it would be embarrassing for an Appointments Committee to decide whether close colleagues—often close friends—should be provided with post-retirement facilities. University funding is, in any case, out of the question because of the general stringency of University finances, and the allocation of funds to retired people could give rise to resentment and criticism from younger claimants. Post-retirement funding has to come from the Research Councils and private grant-awarding foundations. Nor does the University itself provide accommodation although individual Departments can offer space if they happen to have any free.

I was anxious not to retire into professional inactivity because I felt that I was fit, that I was not short of ideas for research, and was still motivated by an insatiable curiosity to explore the unknown.

Fortunately, before the time for retirement came, I began to receive enquiries from various academic institutions asking if I would be interested in continuing my work at their establishment after my retirement. The first enquiry came from the Texas Christian University at Fort Worth through the Vice-Chancellor for Research, W. O. Milligan. It was followed by others from Duke University at Durham, North Carolina (Philip Handler), the University of Wisconsin (Philip Cohen), the University of Chicago (George Beadle and Earl Evans),

Columbia University, New York (Stanley Bradley), the University of Pennsylvania (Samuel Gurin and Britton Chance), Dartmouth Medical School, New Hampshire (Clarke Gray), the University of California at Davis (Paul Stumpf), the Instituto Politecnico Nacional, Centre de Investigacion y de Estudios Advanzados at Mexico (Arturo Rosenblueth), as well as from British Centres—Nottingham University (Fred Dainton), Leicester University (Hans Kornberg), and the University of East Anglia (David Davies).

The British enquiries were made on the understanding that the Medical Research Council would continue its financial support while the universities would provide laboratories. I had in fact learned from the Council that, in principle, there were no regulations to prevent their supporting my research after official retirement and after the winding-up of the Metabolic Research Unit. The support would be on a reduced scale as far as the number of collaborators was concerned but I would be free to apply to the Council for funding, whichever centre offered me accommodation.

In the first instance approaches were informal but by 1965 some of the American proposals had taken a more concrete form and Margaret and I decided to make a visit to explore all the possibilities and general circumstances. After this visit I received several attractive offers from American institutions. However, there were many good reasons for my eventual decision to stay at Oxford. First there was the likelihood that by doing so I could keep the nucleus of my research team together; secondly, there would be little or no interruption to our research; and thirdly, Margaret and I would not have to move our home and leave family and friends.

Having the prospect of the Medical Research Council's continuing support and their permission to retain about eight members of staff—academic and technical—and my research equipment (the Council's property), I drafted a letter to the Hebdomadal Council. I showed the draft to my wise old friend, George Pickering, Regius Professor of Medicine and a Member of the Hebdomadal Council at that time, and asked him for advice.

7 June 1965

My dear George,

Enclosed is the draft of a letter which I was contemplating writing when I spoke to you over the telephone. I shall be most grateful to have your comments and advice, especially on the question whether I should send such a letter to Council or whether the matter might be more satisfactorily dealt with in some other way.

Yours as ever,
Hans

Draft letter to Hebdomadal Council

I feel that I ought to put before Council some facts and comments related to my retirement in 1967 because they bear on questions of general policy.

I was recently asked by American colleagues what I intended to do after my retirement and, in particular, what arrangements would be available at Oxford for continuing my research. When it became known that there were not likely to be opportunities at Oxford, six American universities which had this information offered me a professorship, with excellent facilities and salaries up to $35 000 per annum.

I hasten to say that the salary side is of little importance to me. What I am anxious to do is to carry on with research and to remain in contact with research workers, especially younger ones, and with students. To do this I need about 1500 sq. ft. of laboratory space.

Before making a decision, I would like Council to know that I would much prefer to remain at Oxford, if a suitable laboratory can be provided. There would be no need for any financial support for the research from the University as I can secure this from outside sources.

George managed the matter very well. He spoke to the Registrar of the University, Sir Folliott Sandford, and was advised that I should not send the letter but let the matter be dealt with informally. The outcome was that Paul Beeson, Nuffield Professor Clinical Medicine, offered me about 1100 square feet which had become available to his Department. The space consisted mainly of offices and had to be converted. The Nuffield Committee of Medicine allocated an adequate sum (about £15 000) for the conversion and the new laboratory was ready when we had to leave the Department of Biochemistry in the late summer of 1967.

Although grateful, I was a little sad that the space at the

team's disposal would be so limited. I knew that much laboratory space around me was not fully used. I had asked for 1500 square feet and had been given two-thirds of this. It meant that I had to turn away several useful potential collaborators who wished to join me, that I had to dispose of almost all my scientific journals and destroy many papers which could be of later historical interest. I felt that, once again, I was going to have to justify myself and hope that in time I would be given the relatively little extra space the team needed to be fully effective. This came about in 1969 when Professor Beeson also made available an adjoining room of about 500 square feet which had been vacated when a new animal house was built. The Geigy Company of Basle provided money for its conversion to a laboratory.

I must record here my deep appreciation to both George Pickering and Paul Beeson for their support, given quietly and without fuss. It is one—a minor one—of many good deeds (good, I hope not only for the individuals concerned but also for the Institution we serve) that these outstanding academic personalities have done behind the scenes, not properly appreciated by wider academic circles, or even within our own University.

The most annoying aspect of my leaving the Department was a great muddle in the arrangements for modifying the laboratory for the use of my successor, Rodney Porter. Porter had been given to understand that the alterations he had asked for would be completed by the time he took office, alterations which would mean that I would have to vacate the laboratory three months or so before the appointed time (30 September). All this I learned in casual conversation with Porter when it was already too late to change the plans, since Porter had already agreed to vacate his present laboratory by the end of September. When I took the matter up with the Registry, it seemed that everyone had told everyone else (except me) and that everyone assumed that I had been told.

So I was made to choose between insisting that I retain my laboratory until the appointed time, which would have been a very bad gesture towards Porter, or to put myself and twenty of my collaborators to a great deal of inconvenience, particularly those who were leaving the team and needed all the time to complete the work they had in hand. Fortunately our new laboratories became available earlier than expected and some members of the team moved in July, but we still had the inconvenience of being separated.

By the late summer of 1967 the new, smaller team—Derek Williamson, Patricia Lund, Len Eggleston, Reg Hems, a secretary, four junior technicians, and myself—had moved into the new premises. With us came our current visitors and research fellows, who included Robert Pitts of Cornell University Medical School and R. L. Veech of Washington D.C.

The general setting of the Department of Medicine struck many familiar chords, reminding me of my early work as a hospital clinician at Freiburg. We were all very happy in our new quarters and our hosts in the Department made us feel welcome. A charming testimony to our relations with the Department came in a note from Robert F. Loeb in December 1972, appended to a Christmas card:

Dear Hans,
It was so good to sit next to Paul Beeson at dinner two weeks ago and I envy his association with you at the Nuffield. Wasn't that a good 'marriage' you made?

Bob

The whole team made it its policy to be relevant to the clinicians, helping practically with biochemical problems and participating in collaborative studies. Before very long such joint research was under way not only with the Department of Medicine but also with the Departments of Surgery, Neurology, Obstetrics, Paediatrics, and Neuropathology. What was particularly satisfying to the team was the realization that their 'academic' research had much more practical relevance than they had expected and was of direct interest to the clinicians in the care of patients. This was a great boost to morale.

The decision of my senior colleagues to remain with me in my retirement was not a simple one for them. The Medical Research Council made it very clear that there was no long-term future for them in my team, and that it would be much wiser to look for a permanent appointment elsewhere. (Only Reg Hems and Len Eggleston had some guarantee from the Medical Research Council of permanent appointments, though not necessarily at Oxford.) The Council's representative pressed this point very hard indeed in personal interviews, but I was delighted to see that he made no impression. Personal loyalties prevailed, strengthened perhaps by confidence—based on past members' experience—that suitable appointments would be forthcoming if and when they were needed. I like to think that there was also a sense of satisfaction in being a member of this particular team.

The original grant support from the Medical Research Council was for three years, then it was renewed for five years and is still continuing in 1980 at the time of writing. For most of the time we have also had grants from the United States National Institutes of Health and the National Institute on Alcohol Abuse and Alcoholism.

Every year, the team has published about a dozen full-length papers which have found much recognition in that they are widely cited. We have trained graduate and post-graduate students and post-doctoral fellows, who have included senior medical scientists from the United States. For several months in 1971–2 three professors (Louis Welt of North Carolina and Yale, Alex Leaf of Harvard, and Franklin Epstein of Yale and Harvard) were with us at the same time, Epstein for the second time in his career. Our short-term visitors, from many countries, have also been trained in biochemical techniques, especially organ perfusion and the use of isolated cells, the freeze-clamping method and many analytical procedures.

In the list of publications (p. 268) the papers [245–347] record the volume and range of work in which I have been involved during my 'retirement'. Not infrequently I have been called upon to write and lecture on historical and philosophical aspects of biochemistry or science—this explains some of the titles in the list. Many other papers (not listed) have been published by members of the team during this period, particularly by Derek Williamson and Pat Lund with their students and collaborators.

During the last ten years I have also interested myself in a problem of society—the ever-increasing rise in juvenile delinquency, which stares us in the face day by day. I have felt that those responsible in the last resort for controlling this problem—the legislature, the judiciary, and academic sociologists—do not pay sufficient attention to biological and medical aspects of human nature. Much has been said and written about the social responsibilities of scientists, all too often with tendentious political overtones, and academics are often accused of living in an ivory tower and not concerning themselves sufficiently with their broader responsibilities towards society. I interpret 'the social responsibilities of scientists' as meaning that the scientist must bring his scientific knowledge to bear on social problems. In other words, I think I ought to look at social problems from the standpoint of science and its methodology, in contrast to other scientists who regard their social responsibility as a matter of bringing their political views to bear on science.

From this background I have lectured and published my views on the biological and medical aspects of asocial conduct, on what I consider root causes of modern juvenile delinquency and on what kinds of preventive action might be taken [290, 332, 345]. I cannot demonstrate that my efforts have as yet prevented one single case of asocial conduct, but they have brought me into contact with like-minded people and organizations, such as the Thomas Jefferson Research Center[34] in Pasadena, California and the Farmington Trust in Oxford. One cannot expect to stamp out rapidly an evil that has developed deep roots over two decades or so. But I have hopes that joint efforts by many like-minded people will eventually prompt governments (whose help is essential) into effective preventive action, above all in the area of education. The documented successes of the Thomas Jefferson Research Center justify such hopes.

In recent years I have sometimes been asked, 'Why do you carry on with work? Why are you not enjoying yourself? Why are you not now doing all the things you wanted to do earlier in life and could not because you were too busy?' I find such questions somewhat irritating. After all, there is work and work. I suspect that many who ask such questions have never experienced the profound enjoyment one can derive from creative work—enjoyment, fun and intellectual satisfaction. I am one of those very lucky people who has always derived deep, lasting satisfaction and pleasure from his work. So I am anxious to carry on as long as I feel my work is still creative and as long as I am permitted to do so.

One of my questioners, knowing that I enjoy music, once argued that I should use the opportunity given by retirement to listen to more and more music. He is a musician and believes there could be no greater pleasure for a music-lover than to listen to good music. I think this view is quite wrong. Even for music-lovers there can be one pleasure still greater—the creating of music. I believe that all artists carry on 'working' as long as they are able to do so. Noël Coward once said, 'Work is fun. There is no fun like work.'

Creative work, I must emphasize, can be in many fields, not only in the arts and sciences. All professions, all crafts, all kinds of occupation and many hobbies offer scope for creativity. Many scientists forced to retire from the laboratory occupy themselves by writing about science or by teaching informally.

I suspect that many of my contemporaries share these feelings on retirement and I could list many from the history of science who remained active in research well into old age. So why should I abandon work to follow the well-meaning advice to 'enjoy myself'? I can think of no greater or more profound enjoyment than that I find in doing the things I do.

19
ENVOI

Looking back over my life I am struck by a sense of wonder and thankfulness that I have been an extraordinarily lucky man. I cannot recall ever having been seriously ill; indeed my school reports confirm that I did not miss a single day's attendance after the age of fourteen. Ever since I first went to school I have been blessed with a variety of keen interests which made life always exciting. In my early teens I already felt sure that I wanted to make a career in medicine, and I looked forward to university life with great expectations which were amply fulfilled. At the post-graduate stage I was supremely lucky to have, in Otto Warburg, a magnificent teacher who set me on the right course as an investigator. By good fortune, too, I published my work on the ornithine cycle of urea synthesis just a few months before Hitler came to power. Hailed as a major discovery, it eased for me the problems raised by my enforced emigration. Incidentally it brought home to me how lucky I was to have always a sense of urgency to get on with the job in hand. Had I worked too leisurely or too often postponed work till the next day, life might have run a different course. So when I was once asked if I had any guiding motto, I replied, with this experience in mind, 'Never put off till tomorrow what you can do today'.

I was extraordinarily lucky that it was Britain (through Hopkins) that offered me a new home, for I was very much attracted to the British way of life. I was given the opportunity, in 1965, of expressing formally my attachment and gratitude, when I was asked to act as spokesman for refugees to Britain.[35]

After arriving in England I began a new education in the Cambridge laboratory—a model academic community, working harmoniously for a common purpose in an atmosphere of mutual trust, encouragement, and help, inspired by the outstanding qualities of character and intellect of its central figure.

I became a British citizen within a few days of the outbreak of the Second World War and was thus spared the tribulations of internment[36] which deeply upset many fellow refugees.

As a young researcher I never imagined that a university professorship would be within my reach, and once I had become a professor at Sheffield I never expected to find myself later in a leading position at Oxford. Again I was lucky that the Oxford post became vacant at just the right time for me.

Luck was also on my side in endowing me with an optimistic temperament, which has helped me to look on the pleasant and bright side of life and people; I early found that it is much better to enjoy and encourage the good qualities in people than to worry about and harp on their failings and foibles. It was also a lucky disposition that I found the most satisfaction in trying to do something worth while in research; the high-level posts, which gave me increasingly good research facilities, came as a bonus. I have never had any difficulty in concentrating on my work and was not easily sidetracked from it—for instance by invitations to travel unprofitably or to serve on numerous committees.[37] When I talk about hard work, I mean that I kept at it steadily, not that I sentenced myself to hard labour.

Then I have had the great fortune of a happy family life.[38] We were, and still are, a close-knit family. The children look upon our house as their home. They visit us frequently, and any suggestion from Margaret or myself that we might move to a smaller house is met with expressions of dismay. Margaret and I were once discussing difficulties which friends of ours were having with their grown-up children. 'The basics have gone wrong', said Margaret. 'What do you mean by "the basics"?', I asked. 'To provide a home for the children upon which they can descend at any time and torment you.'

Through my work I have had the opportunity of meeting interesting people from many walks of life and the joy of making many real friends. This found deeply moving expression during my eightieth year when fellow scientists (in the United States at the initiative of Ronald Estabrook and Paul Srere, in Germany at the initiative of Helmut Sies and Hans-Dieter Söling with the support of the German Society for Biological Chemistry, and in Sheffield at the initiative of the British Biochemical Society) organized symposia in my honour in March, May, and July 1980 respectively. These demonstrations of the recognition of my professional efforts were a most pleasurable experience, as has been the world-wide recognition expressed by many honours[39] and by the inclusion of some of my work in school textbooks and curricula.

So I set out expecting nothing special from life—as I had been brought up to do—except to survive by my own efforts. But because I have had more than my share of luck, and a lucky constellation of genes, life has felt special indeed.

NOTES

1. Further evidence of this attitude of Hitler can be found in Albert Speer's book, *Inside the Third Reich* (Weidenfeld and Nicolson, London (1970)), especially pp. 438 and 448.
2. A good historical account of early anti-Semitism is the article by Lucien Wolf in the eleventh edition (1910) of *Encyclopaedia Britannica*. Anti-Semitism in Germany is adequately recorded in the Brockhaus *Enzyclopädie* (1966). A detailed modern sociological study of anti-Semitism is *The authoritarian society* by T. W. Adorno, E. Frenkel-Brunswick, D. J. Levinson, and R. Nevitt-Sandford (Wiley, New York (1974)). This is a general study of prejudice by experienced sociologists. Recent reports and articles appeared in the *Leo Baeck Institute Yearbook*, especially in Vol. 23 (Secker & Warburg, London (1978)) which put the subject into historical perspective.
3. The phrase 'effortless superiority' was coined by H. H. Asquith on 22 July 1908 in a speech at a dinner in the House of Commons. The dinner was given in his honour on the occasion of his appointment as Prime Minister, by the thirty Members of Parliament who, like Asquith, had been educated at Balliol College, Oxford. Asquith's remark was meant to be a humorous reference to the brilliance of Balliol men, suggesting that it was easy for Balliol men to be successful.
4. Much later, when I sought the origin of this proverb, I found that it was a slight distortion of the words, though not of the meaning, of what Apuleius (about AD 125–180) wrote in his *Apologia* (Chapter 43), 'Non ex omni ligno debet Mercurius exsculpi' (You should not carve Mercury from just any wood).
5. These remarks might give the impression that I disapprove of student grants, but this is not so. I am all in favour of giving everyone of ability and inclination the opportunity to benefit from a university education. But I also believe that material help alone is not enough. Many students also need help of an intellectual and spiritual kind if they are to reap the wide benefit that a university offers. Teachers, by their example, must inspire their students to aim at things other than material rewards, to acquire interests and skills that will help them to get involved in creative, deeply satisfying activities and thus to get the most out of life. Students need a sense of motivation which prompts them to do

things that merit doing, irrespective of material gain. This kind of help is of course much more difficult to provide than money.

6. I was surprised by the reference to my 'profound understanding for the psychology of my patients'. I had never realized that my superiors were aware of this aspect of my clinical work because my talks with patients had taken place in strict privacy. I had been taught that taking a detailed case history was very important and I had taken this to heart. I asked innumerable questions and listened patiently and sympathetically to all my patients wished to say, even if it seemed to have no connection with their illness. So they knew I was deeply interested in their wellbeing and this gave them confidence in me and I often found that they were anxious to tell me very much more about their illness than questions could elicit. Some felt it a rare opportunity to confide their innermost worries about marriage, family life, sex, and difficulty with their jobs and in their dealings with people. To some, I found, the physician is father-confessor. I found this close human contact with people from all levels of society not only most interesting but also an invaluable education.

7. 'Suspected' because Wöhler had succeeded in synthesizing urea in the test-tube from ammonium cyanate.

8. Renal biochemistry continued to be a source of interest throughout my life [48, 49, 193–6, 199, 209, 213, 215, 226, 248, 250, 253, 266]. In 1979 the Fifth International Symposium of Biochemical Aspects of Renal Metabolism, at Oxford, was held in my honour. (See *Int. J. Biochem.* **12**, 1–315 (1980) ed. B. Ross and W. G. Guder.)

9. Further reading: Bullock, Alan, *Hitler: a study in tyranny* (Bantam Books, New York (1958); Harper & Rowe (1953)).
Motzkin, Leo, and Comité des Délégations Juives, *Das Schwarzbuch—Tatsachen und Dokumente* (Die Lage der Juden in Deutschland, Paris (1934)).

10. There had been no pogroms in Tunis, though possibly some demonstrations. The writer had believed sensational inventions of the Nazi press.

11. The Prussian Junkers: former landowning nobility of eastern Prussia, from whose families were drawn many members of the officer class and the higher civil service.

12. The Academic Assistance Council was initiated by Sir William (later Lord) Beveridge in May 1933. Early supporters were Professor C. S. Gibson of the Chemistry Department of Guy's Hospital and Professor A. V. Hill. Further active supporters were Sir George Trevelyan, Sir Frederick Kenyon, Dr Walter Adams, Lord Rutherford, Esther Simpson, and Professor F. G. Hopkins.

Later the Academic Assistance Council became 'The Society for the Protection of Science and Learning'. The story of the rallying and support for displaced scholars has been told by Lord Beveridge in his *The defence of free learning* (Oxford University Press, London (1959)).

13. *This English language* (Longman, Green & Co. (1939)) is a very unusual book. E. Denison Ross conceived it primarily as an experiment in language teaching and explained it as follows:

> It has long appeared to me that in the case of all living languages a large and important element essential to their complete mastery has been almost wholly neglected. This element . . . may be said to comprise that natural store of quotation and allusion which every educated individual acquires in his own language, both by conscious study and by unconscious assimilation from his childhood onwards, and from which he draws in order to lend colour and to give emphasis to the spoken and the written word. This stock-in-trade may be said to form the physiognomy of language, as grammar and vocabulary form its anatomy. Just as many families have household words and expressions of their own invention which are meaningless to outsiders, so has every nation a fund of allusion which is often unintelligible to foreigners. This Guide . . . is, I believe, the first attempt to bring together in one book the most popular anecdotes, allusions and quotations peculiar to the English . . . the kind of knowledge that the composer of 'The Times' crossword takes for granted in his readers.

I mention the usefulness of this book because the publication of an updated version seems to me well worth-while. Many immigrants might benefit from it.

14. The following references supply further information about Hopkins's personality:

 Pirie, N. W. A reminiscence of Sir Frederick Gowland Hopkins (1861–1947). *Trends in Biochemical Sciences* **4**, No. 4, April 1979.

 Dale, Sir Henry, *The Royal Society Obituary Notice* **17**, 115 (1948).

 Needham, J. and Baldwin, E. (eds.) *Hopkins and biochemistry* (several articles), Heffer, Cambridge (1949).

 Papers given at centenary meeting (*Proc. R. Soc.* **B156**, 289 (1962)).

15. The Cayman Turtle Farm Limited is located in Grand Cayman, the largest of the three islands that form the British Crown Colony of the Cayman Islands. Commercial green turtle farming was started in 1968, and ten years later the turtle population was over 70 000.

Turtle farming is a relatively new industry. Turtles lay many eggs but normally only a few eggs and young turtles escape their predators. The farm on the Cayman Islands has breeder turtles whose eggs are incubated and, when hatched, the young are kept in nurseries in the sea and fed artificially. Much research had to be done to establish an optimal diet. After four years the meat and carapace (tortoiseshell) are highly valued products. Wild turtles are now protected animals.

16. In this Chapter I have used synonymously the terms 'lactate' and 'lactic acid', and the analogous terms for other acids, such as 'citrate' and 'citric acid'. The ending '-ate' denotes the dissociated acid. The dissociated ($R.COO^-$) and undissociated ($R.COOH$) groups are readily interchangeable. Most, but not all, of the acid molecules are dissociated at the physiological pH. In the text, 'lactate' is used because it is shorter; in the formulae, the undissociated form is used because it avoids the symbol for the negative charge.

17. Oxford University has arrangements whereby exceptionally able technicians can obtain a degree without having attended formal courses. The degree of B.Sc. can be awarded on the basis of a research thesis and an oral examination. Opportunities for attending lectures and evening classes are available but success much depends on a great deal of disciplined homework. The number of technicians who achieve success in this way is very small.

18. See papers [102–59]. Not all the papers published by the team are included in my list of publications (p. 268) because much of the work was done essentially independent of myself.

19. Thirty-two years later, in 1980, the Medical Faculty of the University of Göttingen referred to these events in a document conferring an honorary doctorate upon me. The document, in Latin, after listing my major discoveries, referred to me as a man

'qui tyrannorum vi e patria fugere coactus tamen humanissime collegas Germanicos sine ira iuvit iterum eis aperiens aditum ad communitatem biochemicorum totius orbis'.

[who, by the oppression of tyrants, was forced to flee his homeland, but nevertheless, without vindictiveness had the humanity [at the end of the war] to help his German colleagues to re-enter the worldwide community of biochemists.]

20. This is the subject of 'prochirality'. Chirality or 'handedness' (χεῖρ, hand) is mirror-image symmetry, as shown, for example, by left and right hands. Symmetrical shapes are superimposable, but left and right hands, although they have a similar structure, are not superimposable. They are mirror-images of each other.

Prochirality refers to substances which under certain conditions exhibit the phenomenon of chirality, e.g. citric acid when it is rigidly attached to an asymmetric enzyme.
21. The following paragraphs are extracts from a talk which I gave in 1967, later published in *Nature* [227].
22. If frequent and close contact with fellow scientists is of the utmost importance to creativity, it is no less essential that the scientist also has frequent periods of solitude where he can digest information, reflect upon it, and try out new ideas in depth.

I find that Leonardo da Vinci's comments on the importance of solitude for the artist also holds true for the scientist:

> In order that the wellbeing of the body may not sap that of the mind the artist ought to remain solitary, especially when intent on those studies and reflections of things which continually appear before his eyes and furnish material to be well kept in the memory. While you are alone you are entirely your own; and if you have but one companion you are but half your own, or even less in proportion to the indiscretion of his conduct.

(*Selections from the notebooks of Leonardo da Vinci*. Ed. Irma A. Richter, p. 216. Oxford University Press (1952).)

23. The 'curious things' to which Sinclair refers had to do with the composition of the Electoral Board for Professorships at Oxford. At that time it consisted of seven people, the majority of whom had no detailed information about the candidate's subject and were therefore unable to form a first-hand opinion of his merits. The Board consisted of the Vice-Chancellor, the Head of the college to which the Chair was attached, and a second representative of the college in a similar field to the candidate's, but often still quite remote from it. In the case of the Biochemistry Chair, the other four members were the Regius Professor of Medicine (a clinician), the Professor of Physiology, and two outside experts— Professor Chibnall of Cambridge and Sir Edward Mellanby, retired Secretary of the Medical Research Council. The last two were the only experts in biochemistry. On the whole, the Electoral Boards have done their jobs very satisfactorily, but all depends upon whether one or two members are prepared to take a great amount of trouble to explore the field by consulting all and sundry. From 1955 until his retirement from the Regius Professorship of Medicine in 1968, Sir George Pickering handled this task very effectively.
24. The term 'college' has a special connotation at Oxford. The organization of Oxford University differs radically from all others except Cambridge, and can only be understood on the basis of its

long history. Its precise origins are not known; all we do know is that in the twelfth century scholars began to congregate in Oxford. There was no 'Founder'. The University was a private association. To begin with, teaching took place in ordinary houses but eventually the teachers (mostly clerics) and students lived together in special buildings. This was the start of the 'colleges'. They were small, residential, mainly ecclesiastical schools, and the sum of these schools constituted the University.

The colleges, which together still constitute the University, are autonomous, each with a Governing Body consisting of its Fellows. Like the University, each college is a corporation established by Royal Charter, with the right to make its own laws, within limits. Colleges do not receive Government funds directly; their running expenses are largely defrayed by student fees, but since these nowadays derive almost exclusively from public funds, the colleges are indirectly dependent on public money.

To become a member of the University, a student has to be accepted by a college, and every student and teacher must be attached to a college. Thus the colleges alone control the admission of students to the University as opposed to Faculty Boards or Departments elsewhere. It is roughly true to say that the functions of the colleges are teaching and research, while the University looks after those aspects which demand uniformity of treatment, such as examining for and conferring degrees, and it is responsible for the libraries, laboratories, and museums.

In the sciences, lectures and classes take place in the University science departments. In addition, students receive tutorial teaching within their colleges, each student spending an hour once a week with his tutor for a free discussion of his work. The backbone of these tutorials is usually an essay set by the tutor and discussed during the 'tutorial'. This close personal contact with a tutor is of tremendous value to the student. In the humanities, college teaching, both lectures and tutorials, plays a much greater part than in the sciences where laboratories and demonstration facilities are required.

It is this double teaching system—university courses plus college tutorials—which is the unique feature of Oxford (and Cambridge).

A college also constitutes an important social centre for its members—teachers and students. College life abounds with extracurricular activities of many kinds and there are clubs and societies covering a wide range of interests and hobbies. Colleges are deliberately small—a few hundred students—so that everybody more or less knows everybody else. This is ensured by the

NOTES

rule that every undergraduate must live under the college's roof for at least the first year.

For further details see *Handbook to the University of Oxford*, published by the Clarendon Press; *A History of the University of Oxford* by Sir Charles Mallet published by Methuen; *The Clarendon Guide to Oxford* by A. R. Woolley, published by Oxford University Press.

25. 'Demonstrator' became 'Lecturer' in 1964 to bring Oxford into line with other universities, where 'Demonstrator' usually refers to more junior teaching appointments.
26. *Oxford emoluments (available to full-time salaried staff)*
    College salaries for teaching and administration;
    examination fees (College and University);
    supervisory fees of graduates;
    extra supervisory fees of graduates;
    child allowances;
    health insurance, BUPA;
    children's education allowance;
    housing allowances or free or cheap rents in college houses;
    free or subsidized meals;
    extra University appointments, e.g. Proctor, Safety Officer;
    lecture fees for lectures given outside one's own Department or Faculty.

    This has developed because final decisions are taken by those who are interested parties, ignoring a basic principle of British committee work. (K. C. Wheare, *Government by committee*, Oxford University Press (1955).)
27. Regular consultation between the Heads of the Science Departments was very necessary for the successful operation of the Departments, but the Heads as a group were not recognized as an 'official body' by the University.
28. The *Oxford Magazine* was a valuable weekly journal, written by dons for dons. It ceased publication because of lack of funds.
29. 'Greats' is a four-year course of classical Greek and Latin language and literature, and philosophy.
30. The Commission of Inquiry under Lord Franks listed 170 recommendations in its report. Very few of these have been fully implemented. Recommendation 44, that 'Oxford should take steps to obtain more professorial posts' is one of the few which were acted upon, and the number of Chairs has increased from 104 to 133. The new ones include several Readership upgradings and a Visiting Professorship. About a third of the new Chairs were established by funds from external sources, including eight of the thirteen new science and medical Chairs.

31. The term 'unripe time' was coined by Francis Cornford (see p. 214).
32. The benefactors for the professorships were as follows

Cardiovascular Medicine	British Heart Foundation (appeal)
⎰ Clinical Neurology ⎱ Paediatrics	National Fund for Research into Poliomyeletis and other crippling diseases
Chemical Pathology	E. P. Abraham Research Fund
Clinical Pharmacology	Rhodes Trust
Psychiatry	W. A. Handley Charity Trust
Radiology	Kodak Co.
Social and Community Medicine	Nuffield Provincial Hospitals Trust

33. The glyoxylate cycle does not play a major part in the metabolism of higher animals. The absence of the cycle is the reason why animals cannot convert fat into carbohydrate. Recently the occurrence of the glyoxylate cycle has been demonstrated in toad urinary bladder epithelium. This is the first demonstration of this cycle in higher animals. For some time it has been known to occur in protozoon Tetrahymena (Goodman, D. B. P., Davis, W. L., and Jones, R. G. Glyoxylate cycle in toad urinary bladder: possible stimulation by aldosterone. *Proc. Nat. Acad. Sci. USA* 77, 1521–5 (1980)).
34. Thomas Jefferson Research Center, 1143 North Lake Avenue, Pasadena, California 91104, USA.
35. In 1965 a refugee organization (the Association of Jewish Refugees in Great Britain) proposed that refugees from Nazism and Fascism who had found a home in Britain should express their gratitude by collecting a sum of money to be given to a British institution. Some £90 000 was subscribed and offered to the British Academy to be used for the furtherance of scholarship. I was chosen to present this 'Thanks Offering to Britain' to the Officers of the British Academy (Lord Robbins and Sir Mortimer Wheeler) at a formal ceremony on 8 November 1965, held in the Livery Hall of the Saddlers' Company. Many fellow refugees who listened to or read the speech have said that the feelings I expressed very much reflected their own.

> To me has fallen the privilege of handing over a cheque to the President of the British Academy. This cheque and the efforts leading up to it are no more than a token, a small token of the deep sense of indebtedness which is harboured by all of us who came to this country as refugees and

were given here a new home—not merely a shelter, but a true home. We all went through the experience that home is not always where one was born and brought up. Home is where one strikes roots, where one has the opportunity of doing the things which, by virtue of inclination or conscience or some other forces deep down in one's soul, one feels on ought to do in order to fulfil one's life and thereby gain true happiness. The social climate and the social soil of this country, thanks to the spirit of generosity and tolerance that pervades it, made it easy for us to strike roots and to become firmly settled.

No sum of money can adequately and appropriately express our gratefulness to the British people. Perhaps the only proper way for us to try and repay the debt is to make a continuous effort to be useful citizens, doing a job to the best of our abilities, taking an active part in the general life of the community, fully identifying ourselves with the communal life of the country, and offering our services whenever the occasion arises. If, in the course of trying to serve the community, we have also served ourselves, and done very well for ourselves in many cases—sometimes even embarrassingly well—this is perhaps in the nature of the circumstances, and in particular an outcome of the fairness with which we have been treated.

Quite a large number, of course, served with the Armed Forces. I believe, although statistics are not readily at hand, something like one in every eight refugees here during the war years served, and quite a few paid with their lives; this we should remember today.

I would like to add one further comment. If it was force of circumstances and not our own choice which drove us out of the country of our birth, it was in many cases our free choice to take refuge and to settle in this country rather than in other parts of the globe. No doubt there were many different reasons which prompted individuals to make this choice. I cannot pretend to know all the reasons, but I do know some of my own, and I know that these are shared by very many.

What this country of our adoption gave us was not just a new home and livelihood. What we also found was a new and better way of life, a society whose attitudes to life were in many ways very different from what we had been accustomed to, and, I dare say, accustomed to not only under the Nazi rule. Coming from an atmosphere of political oppression and persecution, of hate and violence, of lawlessness, blackmail and of intrigue, we found here a spirit of friendliness, humanity, tolerance and fairness. We found a society where people of many different dispositions, races, convictions and abilities lived together harmoniously, and yet vigorously. We saw them argue without quarrelling, quarrel without suspecting, suspect without abusing, criticise without vilifying or ridiculing, praise without flattering, being vehement without being brutal. We saw what Robert Browning said of his dog, 'strength without violence, courage without ferocity'. These are some of the characteristics of the soul of this country.

It is the widespread occurrence of these traits which impressed us as being so different from the world from which we had escaped. Of course, all societies have their outstanding individual heroes, but these have not necessarily left their mark on the character of their society as a whole. If proof were needed that these attitudes which I have mentioned are prevalent traits of the British way of life, I would say: which other language uses in its everyday life phrases equivalent to 'fair play', 'gentleman's agreement', 'benefit of the doubt', 'give him a chance', 'understatement', phrases indicative of a sense of justice, of a sense of perspective, of tolerance, of humility and, above all, of respect for humanity? Quite a few languages have to use the English words when wishing to express the sentiments lying behind these phrases, and I am sure this is not an accident.

It is this way of life with which some of us, I for one, fell in love. Others may emphasise different aspects of our society—*our*, I am proud to say—with which they became enchanted and fell in love, and as is usual in love affairs, our affection may have blinded us perhaps to any blemishes the critics might detect. They may argue that our society is not that complete paragon of human affection I have depicted. But still, we do know why we love it. For whatever individual reasons different people came to settle in this country, the theme of their motives was similar, and what I would like to convey is the depth of our feeling for this country of our adoption. It is a very small token of our gratitude which I ask you, Lord Robbins, in your capacity as President of the British Academy, to accept.

36. The tribulations arising from the 1940 internment policy have been documented by Peter and Leni Gillman in *Collar the lot. How Britain interned and expelled its wartime refugees* (Quartet Books, London (1980)).

37. I believe, however, that I did my fair share of committee work. I was closely associated with the Agricultural Research Council for twenty years, during eleven of which I served as a member of Council. I served for five years on the Hebdomadal Council of Oxford University, where Faculty Board meetings were a routine obligation, as were the Senate and Faculty Board meetings at Sheffield. I served on the Council and on committees of the Royal Society, and for two three-year periods I served on the Committee of the Biochemical Society.

38. The bringing up of children whose father has a reputation of distinction is liable to pose special problems. While there may be times when the children feel proud of their father, they may also come to suffer from a sense of inferiority. Margaret, instinctively, coped with this possibility very efficiently so that, I think, none of our children ever suffered in this way. In our home life it was Margaret's policy to play down the importance of my professional standing and to emphasize that in innumerable ways I was quite ordinary, a person who did and said many stupid things and who was, in many areas of human activity,

quite undistinguished, even incompetent. Sometimes she felt that this was rather hard on me, but she knew that I could take it. She also thought, and I agreed, that it was good for me and would help me to keep a sense of proportion in the face of being treated all too often, publicly and professionally, as some kind of hero. For this I owe her a special debt. If our family life and the development of our children have been very happy, it has been due to a large extent to Margaret's loving, sound, and sensible devotion.

Perhaps the following story is appropriate here. Max Born once told me that his son Gustav, as a young man, before Gustav himself became famous as one of the leaders in pharmacology, was occasionally mildly irritated by people asking, 'Are you the son of Max Born?'. The irritation disappeared after he had named his first son 'Max', in the family tradition of naming children after their grandparents. After that, when asked 'Are you the son of Max Born?' he replied with a proud smile, 'I am his father'.

I mentioned in Chapter 15 that our younger son, John, was now a behavioural zoologist. All three children studied science at Oxford. Paul is now a systems analyst in industry and Helen is a botanist (although at present she is a full-time mother). We have four grandchildren.

39. I record the honours which have been conferred upon me—although it embarrasses me because it sounds like boasting. I do it to show that scientific research receives plenty of recognition, sometimes perhaps undeserved recognition. Robert K. Merton, the distinguished American sociologist, had drawn attention (*Science, N.Y.* **159**, 56 (1968)) to the phenomenon that scientists who are firmly established, either by having made one major contribution to their subject or by holding an important post, receive more recognition than they deserve, in terms of invitations to lecture, honorary degrees, prizes, and being quoted in the literature as an authority. When they publish a paper jointly with an unknown junior collaborator, an inordinate amount of credit is liable to go to the well-established name even if he is not the main author. Merton calls this allocation of credit to those who are already famous the 'Matthew Effect' in the reward system: 'For unto everyone that hath shall be given and he shall have abundance; but from him that hath not shall be taken away even that which he hath.' (Matt. 25:29).

General

Knight Bachelor
Nobel Laureate
Fellow of The Royal Society of London

Honorary degrees

D.Sc. Chicago, Philadelphia, Sheffield, Leeds, London, Leicester, Bristol, Wales, Liverpool, Indiana
Sc.D. Cambridge
Ll.D. Glasgow
Ph.D. Jerusalem
M.D. Freiburg, Hanover, Berlin (Humboldt), Göttingen
Doctor Paris, Bordeaux, Valencia, Granada

Honorary and foreign membership of learned and professional societies

(1951) Nazionale Accademia dei Lincei, Rome
Socio Straniero
(1952) Académie Nationale de Médecine, Paris
Correspondant Etranger
(1955) National Institute of Sciences of India
Honorary Fellow
(1957) American Academy of Arts and Sciences
Foreign Honorary Member
(1958) Royal College of Physicians
Fellow
(1960) American Philosophical Society
Member
Biochemical Society of Israel
Honorary Member
(1961) American Society of Biological Chemists
Honorary Member
(1962) Académie Royale de Médecine de Belgique
Membre Honoraire Étranger
(1964) US National Academy of Science
Foreign Associate
Consejo Superior de Investigaciones Cientificas
Consejero de Honor
(1967) Royal College of Pathologists
Honorary Fellow

Biochemical Society
  Honorary Member
Trinity College, Oxford
  Honorary Fellow
(1968) Royal Society of Medicine (Section of Experimental Medicine and Therapeutics)
  Honorary Member
St Catherine's College, Oxford
  Honorary Fellow
(1969) Indian Academy of Science
  Honorary Fellow
Institute of History of Medicine and Medical Research, New Delhi
  Honorary Fellow
Sociedad Española de Bioquimica
  Socio de Honor
Deutsche Akademie der Naturforscher Leopoldina
  Ehrenmitglied (Mitglied since 1956)
Istituto Lombardo, Milan
  Membre Straniero
(1972) American Association of Physicians
  Honorary Member
Deutsche Gesellschaft für Innere Medizin
  Ehrenmitglied
Weizmann Institute of Science
  Honorary Fellow
(1973) Societé de Biologie, Paris
  Membre d'Honneur
Académie Nationale de Médecine, Paris
  Associé Étranger
Physiological Society
  Honorary Member
Order pour le mérite, Germany
  Foreign Member
(1974) Bayerische Akademie der Wissenschaften (Mathematisch-Naturwissenschaftliche Klasse)
  Korrespondierendes Mitglied
(1975) Akademie der Wissenschaften in Göttingen (Mathematisch-Physikalische Klasse)
  Korrespondierendes Mitglied
(1976) Royal Society of Edinburgh
  Fellow
Società Italiana per il Progresso delle Scienze
  Socio d'Onore

(1977)   Académia de Ciencias Medicas, Córdoba, Argentina
Académico Correspondiente Extranjero
(1978)   Institute of Biology
Honorary Fellow
(1980)   Society for General Microbiology
Honorary Member.

Awards

Honorary Citizen of Hildesheim, Germany
Royal Medal, Royal Society
Copley Medal, Royal Society
Gold Medal, Netherlands Society for Physics, Medical Science and Surgery
Gold Medal, Royal Society of Medicine
Otto Warburg Medal, Deutsche Gesellschaft für biologische Chemie
Gold Medal of the City of Paris, Comité National de Biochimie
Lasker Award, American Public Health Association
Purkinje Gold Medal, Czechoslovak Academy of Science
Gold Medal, University of Bari
Award of the American College of Physicians, Boston.

A few rather amusing 'honours' have been bestowed upon me in the United States:
By the authority of the State of Texas I was commissioned Admiral in the Texas Navy. I am told this navy has over 1000 Admirals, but no ship afloat—only one old battleship moored in a creek.

In 1966 I was officially appointed to the office of Alcalde (Mayor) of the city of La Villita (the original San Antonio) in Texas.

I became a Chieftain of an Indian Tribe when Indiana's Governor, Roger D. Branigin, appointed me 'Sagamore of the Wabash', of which tribe there are no members but many chieftains (sagamores).

PEOPLE IN THE BOOK

Abderhalden, Emil (1877–1950). German biochemist of Swiss birth, Berlin, Halle. Pupil of Emil Fischer.
Acheson, R. Morrin (b. 1925). British chemist, Oxford. Lecturer in Department of Biochemistry.
Ackermann, D. (1878–1965). German biochemist, Würzburg.
Acland, Sir Richard (Thomas Dyke) (b. 1906). British politician.
Adams, Sir Walter (b. 1906). British historian, educator, Salisbury (Rhodesia), London.
Adrian, Edgar Douglas (later Lord Adrian) (1889–1977). British neurophysiologist, Cambridge.
Ajl, Samuel Jacob (b. 1923). American biochemist, Ames (Iowa), Washington.
Alder, Kurt (1902–58). German chemist, Cologne University.
Ammon, Robert (b. 1902). German biochemist, Berlin, Königsberg, Homburg (Saar).
Ampère, André Marie (1775–1836). French physicist and mathematician.
Anson, Mortimer Louis (1901–1968). US biochemist, Princeton, New York.
Aschoff, Eva. German bookbinder, daughter of Ludwig Aschoff.
Aschoff, Ludwig (1866–1942). German Professor of Pathology, Freiburg.
Asquith, Herbert H. (1852–1928). British politician, Prime Minister 1908–1916.
Attlee, Clement R. (1883–1967). British politician, Prime Minister 1945–1951.
Aub, Joseph C. (1890–1973). US medical scientist, Harvard Medical School.
Auburn (Auerbach), Walter (1906–1979). German/British Physician, Altona, Manchester, Auckland (New Zealand).
Austen, Canon Arthur. Chaplain in British armed forces during Second World War, Vicar in Sheffield.
Austen, Dorothy. My secretary 1945–1949. Wife of above.

Bach, Stefan J. German/British chemist/biochemist, Cambridge, Bristol.
Bacon, John S. D. (b. 1917) British biochemist, Cambridge, Sheffield, Aberdeen. Member of Biochemistry Dept., Sheffield during HAK's chairmanship.

von Baeyer, Adolf (1835–1917). German organic chemist, Munich.
Baldwin, Ernest H. F. (1909–1969). British biochemist, Cambridge, London.
Barcroft, Joseph (1872–1947). British physiologist, Cambridge.
Bartley, Walter (b. 1916). British biochemist, Sheffield. My collaborator in Sheffield and Oxford.
Beadle, George Wells (b. 1903). US geneticist, California, Chicago.
Beckman, Arnold O. (b. 1900). US instrument maker, Fullerton, California.
Beeson, Paul (b. 1908). US physician, Professor of Medicine, Yale, Oxford.
Bennett, George McDonald (1892–1959). British chemist, Sheffield, London.
Benzinger, Theodor Hannes. German/American physiologist, Göttingen, Bethesda (Maryland). HAK's student in Freiburg 1932 and collaborator in 1950s.
Bernal, John Desmond (1901–1971). British X-ray crystallographer and writer.
Berry, Michael Nathaniel (b. 1934). British biochemist, Adelaide (Australia).
Berthollet, Claude Louis (1748–1822). French chemist.
Beveridge, William (Lord) (1879–1963). British barrister, economist, educator, and Government adviser. London School of Economics, Master of University College, Oxford.
Bielschowsky, Franz. German/British physician biochemist, Freiburg, Madrid, Sheffield, Dunedin. My colleague in Freiburg and Sheffield.
Bismarck, Otto (1815–98). German statesman.
Blackett, Patrick Maynard Stuart (Lord) (1897–1974). British physicist, President of the Royal Society (1967–70), Birkbeck, Manchester, Imperial College, London.
Blaschko, Hermann (Hugh) (b. 1900). German/British biochemical pharmacologist, Berlin, Heidelberg, London, Cambridge, Oxford. Personal friend since 1919.
Booth, Vernon (b. 1900). British biochemist, Cambridge. Personal friend since 1933.
Born, Gustav (b. 1921). British pharmacologist, Oxford, Cambridge, London.
Born, Max (1882–1970). German/British biochemist, Göttingen, Cambridge, Edinburgh.
Bowra, Sir Maurice (1898–1971). British classicist, Oxford.
Bradley, Stanley. US physician, Columbia University (New York).
Bragg, Sir William (Lawrence) (1890–1971). British physicist, Manchester, Cambridge, London.

PEOPLE IN THE BOOK

Braunstein, Alexander E. (b. 1902). Russian biochemist, Moscow.
Bronk, Detlev (1897–1975). US biophysicist and academic administrator, Philadelphia, Johns Hopkins, Rockefeller University.
Browning, Robert (1812–1889). British poet.
Bücher, Theodor (b. 1914). German biochemist, Marburg, Munich. Collaborator of Otto Warburg.
Buchner, Edward (1860–1917). German chemist, Berlin.
Buck, Paul. USA, Provost of Harvard University.
Burk, Dean (b. 1904). US biochemist, Washington, Bethesda.
Burton, Kenneth (b. 1926). British biochemist, Sheffield, Oxford, Newcastle-on-Tyne. My collaborator in Sheffield and Oxford.
Butenandt, Adolf (b. 1903). German biochemist, Danzig, Berlin, Tübingen, Munich.

Callow, Ann Barbara. British biochemist, Cambridge. Librarian in the Cambridge Biochemical Laboratory.
Carter, Cyril (1897–1969). British biochemist, Reader in Biochemistry, Oxford. The senior academic staff member at the time of my arrival in Oxford.
Cecil, Rupert (b. 1917). British, Senior Research Officer, Biochemistry Department, Oxford.
Chain, Sir Ernst (Boris) (1906–79). German/British biochemist, Cambridge, Oxford, Rome, London.
Chamberlain, Houston Stewart (1855–1927). British born writer and 'Kulturphilosoph'. Lived in Germany and Austria. Married 1908 Richard Wagner's daughter Eva.
Chamberlain, Neville (1869–1940). British politician. Prime Minister at outbreak of Second World War.
Chance, Britton (b. 1913). US biochemist and biophysicist.
Chargaff, Erwin (b. 1905). Austrian born US biochemist New York.
Cherwell, (1st Viscount), Frederick Alexander Lindemann (1886–1957). British physicist, Oxford. Friend and adviser of Sir Winston Churchill.
Chevreuil, Michel Eugène (1786–1889). French chemist.
Chibnall, Albert Charles (b. 1894) British biochemist, London, Cambridge.
Chilver, G. A. F. (b. 1910) Oxford Classics Don and Professor at University of Kent.
Christian, Walter (1907–55). German biochemist. Longstanding technician and collaborator with Otto Warburg.
Churchill, Lady (Clementine Spencer) (1885–1978)
Churchill, Sir Winston (1874–1965). British statesman.
Clark, George Albert (1894–1963). British physiologist, Sheffield.

Cohen, Philip P. (b. 1908). US biochemist, Madison (Wisconsin). My post-doctoral student at Sheffield.
Cohen, Robert, H. C. Senior officer of the Medical Research Council.
Conant, James Bryant (1893–1978). US chemist, President of Harvard University.
Cori, Carl Ferdinand (b. 1896). Austrian/US biochemist, Buffalo, Washington University (St Louis, Missouri), Harvard Medical School.
Cori, Gerty T. (1896–1957). Austrian/US biochemist, Buffalo, Washington University (St Louis, Missouri), Wife of above.
Cornford, Francis (1874–1943). British classical scholar, Cambridge.
Cornforth, John W. (b. 1917). British chemist, Oxford, University of Sussex (Brighton).
Correns, Carl Erich (1864–1933). German plant geneticist, Leipzig, Münster, Dahlem. One of the re-discoverers of Mendel's Law.
Coward, Sir Noël (1899–1973). British playwright and actor.
Cremer, Werner. Co-worker of Otto Warburg around 1930.
Crick, Francis Harry Compton. (b. 1916) British biochemist, Cambridge, San Diego.
Crowther, James Gerald (b. 1889). British writer on history and social relations of science, London.
Cuvier (Baron), Leopold Chrétien Frédéric Dagobert (Georges) (1769–1832). French anatomist and palaeontologist. Chancellor of University, Paris.

Dainton, Sir Frederick (Sydney) (b. 1914). British physical chemist and scientific administrator, Cambridge, Leeds, Nottingham, Oxford, Chairman, University Grants Committee 1973–78.
Dakin, H. D. (1880–1952). British/US biochemist, Leeds, London, New York.
Dalziel, Keith (b. 1921). British biochemist, Sheffield, Oxford.
Davies, David D. British plant physiologist, London, Norwich.
Davies, Robert Ernest (b. 1919). British biochemist, Sheffield, Philadelphia.
Davy, Sir Humphry (1778–1829). British chemist, Royal Institution, London.
Delbrück, Max (1906–1980). German/US physicist and biophysicist, Göttingen, Pasadena.
Deuticke, Hans-Joachim, German biochemist.
Dickens, Frank (b. 1899). British biochemist, Middlesex Hospital.
Diels, Otto Paul Hermann (1876–1954). German biochemist, Kiel.
Dixon, Gordon Henry (b. 1930). British biochemist, Cambridge, Seattle, Vancouver, University of Sussex, Calgary.

PEOPLE IN THE BOOK 251

Dixon, Kendal Cartwright (b. 1911). British biochemist, Cambridge.
Dixon, Malcolm (b. 1899). British biochemist, Cambridge.
Donegan, Joseph F. Irish physiologist, Galway.
Donnan, Frederick George (1870–1956). British physical chemist, Liverpool, London.
Drury, Sir Alan (Nigel) (b. 1899). British pathologist, Cambridge, London.
Dühring, Carl Eugen (1833–1921). German political writer and philosopher.

Ebert, Friedrich (1871–1925). German politician. First President of Weimar Republic.
Edson, Norman Lowther (1904–1970). New Zealand biochemist, Otago. My collaborator at Cambridge.
Eggleston, Leonard Victor (1920–1974). My longstanding collaborator at Sheffield and Oxford.
Einstein, Albert, (1879–1955).
Eisenhower, Dwight David (1890–1969). 34th President of USA.
Eitel, Hermann (1868–1944). Administrative Director of the University Hospitals, Freiburg.
Eitel (son of above), Hermann (b. 1902). German surgeon, Freiburg-Aalen. My colleague and collaborator.
Elliott, Kenneth Allan Caldwell (b. 1903). South African born Canadian biochemist, Cambridge, Montreal.
Embden, Gustav (1874–1933). German biochemist, Frankfurt.
Emerson, Robert (1903–1959). US plant physiologist, Dahlem, Harvard Urbana.
Epstein, Franklin Harold (b. 1924). US physician, Boston.
Estabrook, Ronald W. (b. 1926). US biochemist, Philadelphia, Dallas.
Evans, Earl Alison Jr. (b. 1910). US biochemist, Chicago. My collaborator at Sheffield.
Exley, Donald (b. 1922). British biochemist, Sheffield, Oxford, London.

Farrell, C. Rosemary (b. 1949). Technician in my laboratory 1965–present.
Felix, Kurt (1888–1960). German biochemist, Munich, Frankfurt.
Fieldhouse, Margaret (*see* Krebs, Margaret).
Fischer, Emil Hermann (1852–1919). German organic chemist, Munich, Erlangen, Würzburg, Berlin.
Fischer, Eugen (1874–1967). German anatomist and anthropologist.
Fisher, Herbert, A. L. (1865–1940). British historian, parliamentarian, educator, London, Sheffield, Oxford.
Fisher, R. B. (David). British biochemist, Oxford, Edinburgh.

Florkin, Marcel (1900–79). Belgian biochemist, Liège.
Foster, Sir Michael (1836–1907). British physiologist, London, Cambridge.
Franck, James (1882–1964). German/US physicist, Göttingen, Baltimore, Chicago.
Franks, Oliver Shewell, (Baron) (b. 1905). British philosopher, administrator and diplomat, Glasgow, Oxford, Washington.
Friedmann, Ernst. German-born biochemist, Strasbourg, Berlin, Cambridge.
Fujita, Akiji (b. 1895). Professor of Biochemistry at Kitasato and Kyoto. Student of L. Michaelis and O. Warburg.

Gaffron, Hans (1902–1979). German/US photobiologist, Berlin, Chicago, Tallahassee.
Gascoyne, Anthony (b. 1932). My technician in Department of Biochemistry, Oxford.
Gay-Lussac, Joseph Louis (1778–1850). French chemist and physicist.
Gebauer, Johannes (1868–1951). German historian. My teacher at 'Andreanum', Hildesheim.
Genevois, Louis. French plant physiologist, Dahlem, Bordeaux.
Gerard, Ralph (1900–1974) US physiologist, Dahlem, Ann Arbor (Chicago).
Gibson, C.S. (1884–1950). Professor of Chemistry, London University.
Goebbels, Josef (1897–1945).
Goldscheider, Alfred. German physician, Berlin.
Goldschmidt, Richard (1878–1958). German/US zoologist and geneticist, Berlin, Berkeley (California).
Gollwitzer-Meier, Klothilde (1894–1954). German scientific physician, Greifswald, Frankfurt, Berlin, Hamburg.
Gray, Clarke Thomas (b. 1919). US microbiologist and biochemist, Dartmouth Medical School.
Green, David Ezra (b. 1910). US biochemist, Cambridge, New York, Madison.
Gregory, Philip (b. 1947). Technician in my team at Oxford 1964–77.
Grey, Sir Edward (1862–1933). (Viscount since 1916) British statesman.
Gunsalus, Irwin Clyde (b. 1912). US biochemist, Madison, Urbana.
Gurin, Samuel (b. 1905). US biochemist, Philadelphia, University of Florida.
Gustav Adolf, King of Sweden (1882–1973).

Haas, Erwin (b. 1906). Hungarian-born German/US biochemist, Berlin, Chicago, Cleveland (Mt. Sinai Hospital). Otto Warburg's technician.

PEOPLE IN THE BOOK

Haber, Fritz (1868–1934). German physical chemist, Karlsruhe, Berlin.
Haddow, Sir Alexander (1907–1976). British pathologist, London, Chester Beatty Research Institute.
Hahn, Otto (1879–1968). German chemist and physicist, Berlin.
Haldane, John Burdon Sanderson (1892–1964). British physiologist and geneticist, Cambridge, London.
Hall, Sir Arthur (John) (1866–1951). British physician, Sheffield.
Hamburger, Viktor (b. 1900). German/US zoologist. Dahlem, Freiburg, Chicago, St Louis.
Hammarsten, Einar (1889–1968). Swedish biochemist, Stockholm.
Hämmerling, Joachim (1901–1979) German zoologist, Dahlem, Wilhelmshafen.
Handler, Philip (b. 1917). US biochemist, Duke University (Durham, N. Carolina). President of the National Academy of Sciences.
Hanke, Martin (b. 1898). US biochemist, Chicago.
Happold, Frank Charles (b. 1902). British biochemist, Leeds.
Harington, Charles (1897–1972). British biochemist, Edinburgh, London.
Harris, Leslie (b. 1898). British biochemist, Cambridge.
Harrison, Douglas Creese (b. 1901). British biochemist, Sheffield, Belfast.
Harrod, Sir Roy (b. 1900). British economist, Oxford.
Hartmann, Max (1876–1962). German zoologist, Dahlem.
Hastings, Albert Baird (b. 1895). US biochemist, New York, Chicago, Harvard, San Diego.
Haworth, Robert Downs (b. 1898). British chemist, Oxford, Newcastle, Sheffield.
Heidegger, Martin (1889–1976). German philosopher, Freiburg.
Hele, Thomas Shirley (1881–1953). British biochemist, Cambridge.
Hems, Reginald (b. 1922). British biochemist. My collaborator at Sheffield and Oxford.
Henseleit, Kurt (1908–1973). German physician. My collaborator at Freiburg. Later Friedrichshafen.
Herkel, Walter (b. 1906). Colleague at Freiburg. Until 1945 physician and teacher of Internal Medicine at Giessen University. Afterwards senior physician at a hospital at Geisenheim/Rhein.
Hess, Benno (b. 1922). German biochemist, Heidelberg, Dortmund.
Hess, Walter Rudolf (1881–1973). Swiss physiologist, Zürich.
de Hevesy, George (1885–1966). Hungarian-born British physical chemist and biochemist, Manchester, Copenhagen, Freiburg, Stockholm.
Heyningen, W. E. ('Kits') van (b. 1911). British biochemist and

microbiologist, Cambridge, London, Oxford. Master of St Cross College, Oxford.
Hill, Archibald Vivian (1886–1977). British physiologist, Cambridge, Manchester, London.
Hill, Robert (Robin) (b. 1899). British biochemist, Cambridge.
Himsworth, Sir Harold (Percival) (b. 1905). British physician and Secretary of Medical Research Council, London.
Hitler, Adolf (1889–1945).
Hodgkin, Dorothy Mary Crawford (b. 1910). British physical chemist, Oxford.
Holmes, Barbara (née Hopkins) (1899–1981). British biochemist, Cambridge. Daughter of Sir F. G. Hopkins.
Holmes, Eric. British biochemist, Cambridge.
Hopkins, Sir Frederick Gowland (1861–1947). British biochemist, Cambridge. President of the Royal Society 1930–1935.
Hoppe-Seyler, Felix Immanuel (1825–1895). German biochemist, Strasbourg.
Horowitz, Norman Harold (b. 1915). US geneticist, Pasadena.
Huggins, Charles Brenton (b. 1901). Canadian/US surgeon and research scientist, Chicago.
Hughes, David E. (b. 1915). British biochemist and microbiologist, Sheffield, Oxford, Cardiff. My collaborator at Sheffield and Oxford.
Hume, E. Margaret. British nutritionist, London.

Ilic, Vera (b. 1941). Technician with my team since 1969.
Iveagh (2nd Earl), Rupert E. C. L. G. (1874–1967).
Iwasaki, Ken (1891–1978). Collaborator of O. Meyerhof. Professor of Biochemistry at University of Kanazawa.

Johnson, William Arthur (b. 1916). My Ph.D. student at Sheffield, later industrial chemist.
Jordan, Arthur (1908–1975). British Clinical biochemist, Sheffield.
Jores, Arthur (b. 1901). German physician, Professor of Internal Medicine, Hamburg.

Kamen, M. US biochemist, Berkeley.
Keilin, David (1887–1963). Polish-born British parasitologist and biochemist, Cambridge.
Kekulé, Friedrich August (1829–96). German chemist, Ghent, Bonn.
Kempner, Walter (b. 1903). German/US physician, Berlin, Durham (N. Carolina).
Kendrew, Sir John (Cowdery) (b. 1917). British biochemist and biophysicist, Cambridge, Heidelberg.

PEOPLE IN THE BOOK

Kent, Paul Welberry (b. 1923). British biochemist, Oxford, Durham.
Kenyon, Sir Frederic (1863–1952). British historian, Director of British Museum, Chairman of the Society for the Protection of Science and Learning, London.
King, Peter P. British Chemical Engineer, Imperial Chemical Industries (Billingham).
Kistiakowsky, George Bogdan (b. 1900). Russian/US chemist, Princeton, Washington, Harvard.
Kitzinger, Charlotte. German scientist. Collaborator of T. H. Benzinger.
Klingenberg, Martin (b. 1928). German biochemist, Marburg Munich.
Knoop, Franz (1875–1946). German biochemist, Freiburg, Tübingen.
Kornberg, Sir Hans (Leo) (b. 1928). British biochemist. My collaborator at Sheffield and Oxford, later Leicester, Cambridge.
Kossel, Albrecht (1853–1927). German biochemist, Heidelberg.
Kraepelin, Emil (1865–1926). German psychiatrist, Munich.
Krebs, Alma (1870–1919). My mother.
Krebs, Elisabeth (b. 1895). My sister.
Krebs, Georg (1867–1939). My father.
Krebs, Gisela (b. 1932). My half-sister. German economist.
Krebs, Helen (b. 1942). My daughter. Botanist.
Krebs, John (b. 1945). My son. Behavioural zoologist, Oxford.
Krebs, Margaret (b. 1913). My wife.
Krebs, Maria (1902–1977). My stepmother.
Krebs, Paul (b. 1939). My son. Computer specialist.
Krebs, Wolfgang (b. 1902). My brother.
Krehl, Ludolf von (1861–1937). German physician. Professor of Internal Medicine, Heidelberg.
Kubowitz, Fritz (b. 1902). Technical co-worker of Otto Warburg, Berlin, Freiburg.
Kuhn, Richard (1900–1967). Austrian/German biochemist and organic chemist, Zürich, Heidelberg.

Lambert, Robert A. Officer of Rockefeller Foundation.
Langley, John Newport (1852–1925). British physiologist, Cambridge.
Laplace, Marquis de, Pierre Simon (1749–1827). French scientist, Paris. President Académie Française.
Lascelles, June (b. 1924). Australian biochemist and microbiologist, Oxford, Los Angeles.
Laue, Max von (1879–1960). German physicist, Berlin.
Lavoisier, Antoine Laurent (1743–94). French chemist, Paris.

Leaf, Alexander (b. 1920). US medical scientist and biochemist, Harvard.
Lehnartz, Emil (b. 1898). German biochemist, Frankfurt, Münster.
Lemberg, Rudolf (1896–1975). German/British biochemist, Heidelberg, Cambridge, Melbourne.
Lichtwitz, Leo. German physician, Göttingen, Altona, Berlin, New York.
Liebig, Justus von (1803–1873). German chemist, Giessen.
Lindsay of Birker (1st Baron), Alexander Dunlop Lindsay (1879–1952). British classicist, Oxford, Master of Balliol College.
Lipmann, Freda. Wife of Fritz Lipmann.
Lipmann, Fritz Albert (1899–1969). German/US biochemist, Boston, New York.
Loeb, Jacques (1859–1924). German/US biologist, Würzburg, New York.
Loeb, Robert Frederick (1895–1973). US physician. Professor of Medicine, Columbia University. Son of above.
Löffler, Wilhelm (1887–1973). Swiss physician. Professor of Internal Medicine, Zürich.
Lohmann, Karl (1898–1978). German biochemist, Berlin.
London, E. S. Russian physiologist, Leningrad.
Londonderry (8th Marquess of), Edward Charles Stewart Robert Vane-Tempest-Stewart (1902–1955).
Lund, Walter Guerrier (b. 1912). British plant physiologist, Windermere.
Lund, Patricia (b. 1933). British biochemist. Member of my team at Oxford.
Lunt, Mary R. (b. 1933). British biochemist, Oxford.
Lynen, Feodor (1911–79). German biochemist, Munich.

McCance, Robert Alexander (b. 1898). British nutritionist, London, Cambridge.
McIlwain, Henry (b. 1912). British biochemist, Sheffield, London.
Mandelstam, Joel (b. 1919). British microbiologist, London, Oxford.
Mangold, Otto, German zoologist, Dahlem.
Marshall, E. K. (1889–1966). US physiologist and pharmacologist, Johns Hopkins University.
Martius, Carl (b. 1906). German biochemist, Tübingen, Würzburg, Zürich.
Masefield, John Edward (1878–1967). British Poet Laureate (1930–1967).
Masson, Sir (James) Irvine (Orme) (1887–1962). British chemist, London, Durham, Sheffield. Vice-Chancellor of Sheffield University.

PEOPLE IN THE BOOK

Medawar, Sir Peter (Brian) (b. 1915). British biologist, Oxford, Birmingham, London.

Meitner, Lise (1878–1968). Austrian physicist, Vienna, Berlin, Stockholm.

Mellanby, Sir Edward (1884–1955). British medical scientist. Secretary of Medical Research Council, London.

Mellanby, Kenneth (b. 1908). British zoologist, Cambridge, Sheffield, Ibadan, Rothamsted Experimental Station, Harpenden, Moulswood Experimental Station, Huntingdon.

Mendel, Bruno (1897–1959). German/Canadian medical scientist, Berlin, Toronto, Amsterdam.

Mendel, Hertha. Wife of Bruno Mendel.

Merton, Robert K. (b. 1910). US sociologist, Columbia University, New York.

Meyer, Karl (b. 1899). German/US biochemist, Dahlem, Zürich, New York.

Meyer, Ovid (b. 1900). Professor of Medicine, Madison (Wisconsin).

Meyerhof, Otto (1884–1951). German biochemist, Kiel, Dahlem, Paris, Philadelphia.

Michaelis, Leonor (1875–1949). German/US biochemist, Berlin, Japan, Johns Hopkins University, Rockefeller Institute.

Miller, Dr. Officer of the Rockefeller Foundation.

Milligan, Winfred Oliver (b. 1908). US chemist, Texas Christian University of Fort Worth.

Mitchell, Peter (b. 1920). British biochemist, Cambridge, Edinburgh, Bodmin (Cornwall). Nobel Prize 1978.

Möllendorff, Wilhelm von (1887–1944). German anatomist, Freiburg, Hamburg, Kiel, Zürich. Supervisor of my M.D. thesis.

Moore, Thomas. British biochemist, Cambridge.

Morgenstern, Christian (1871–1914). German poet and writer—especially of 'nonsense' and of the grotesque.

Morton, R. A. (1899–1977). British biochemist, Liverpool.

Mosley, Sir Oswald (Ernald) (1896–1980). British politician.

Müller, Friedrich von (1858–1941). Professor of Medicine, Munich.

Nachmansohn, David (b. 1899). German/US biochemist and biophysicist, Dahlem, Paris, New York Newhaven. Personal friend since 1924.

Needham, Dorothy Mary Moyle (b. 1896). British biochemist, Cambridge.

Needham, Joseph (b. 1900). British biochemist and sinologist.

Negelein, Erwin (1897–1979). German biochemist, Berlin-Dahlem and East Berlin.

Newsholme, Eric A. (b. 1935). British biochemist, Cambridge, Oxford.
Neuberg, Carl (1877–1956). German biochemist, Berlin, New York.
Notton, Brenda M. (b. 1936). Technician with my team at Oxford.

O'Brien, J. R. P. British biochemist, Oxford.
Ochoa, Severo (1905). Spanish/US biochemist, Dahlem, London, Madrid, St Louis, New York.
Ogston, Alexander George (b. 1911). British biochemist, Oxford, Canberra.
Ord, Margery (b. 1927). British biochemist, Oxford.
Ostern, Pawel (1902–1943). Polish biochemist, Lwow, Freiburg, Cambridge. My collaborator at Freiburg.
Osterwald, Dr. Otolaryngologist, Hildesheim.
Öström, Åke. Swedish biochemist, Stockholm. My collaborator at Sheffield.

Parnas, Jacob (1884–1949). Polish biochemist, Lwow.
Parsons, Denis (b. 1917). British biochemist, Oxford.
Pasternak, Charles (b. 1930). British biochemist, Oxford, London.
Pasteur, Louis (1822–1895). French scientist. Paris.
Peacocke, Arthur Robert (b. 1924). British chemist, biochemist, and theologian, Birmingham, Oxford, Cambridge.
Perutz, Max Ferdinand (b. 1914). Biochemist and biophysicist, Cambridge.
Peter, Albert (1853–1937). German botanist, Göttingen.
Peters, Rudolph (Albert) (b. 1889). British biochemist, Cambridge, Oxford.
Pflüger, Eduard (1829–1910). German physiologist, Bonn.
Phillips, Sir David (Chilton) (b. 1924). British molecular biologist, London, Oxford.
Pickering, Sir George (White) (b. 1904). British physician and medical scientist, London, Oxford.
Pirie, Norman Wingate ('Bill') (b. 1907). British biochemist, Cambridge, Rothamsted.
Pitts, Robert Franklin (b. 1908). US physiologist, Johns Hopkins, New York.
Planck, Max (1858–1957). German physicist, Berlin.
Pohl, Robert (1884–1973). German physicist, Göttingen.
Polanyi, Michael (1891–1976). Hungarian born/German/British scholar. Physical chemist, economist, and philosopher, Berlin, Manchester, Oxford.
Pomerat, Gerard (b. 1901) US biologist, Officer of the Rockefeller Foundation.

PEOPLE IN THE BOOK

Porter, Rodney R. (b. 1917). British biochemist, Cambridge, London, Oxford.
Prelog, Vladimir (b. 1906). Yugoslav/Swiss chemist, Prague, Zagreb, Zurich.
Pringle, John William Sutton (b. 1912). British zoologist, Cambridge, Oxford.

Quastel, Juda Hirsch (b. 1899). British biochemist, Cambridge, Cardiff, Rothamsted, Montreal, Vancouver.
Quayle, John Rodney (b. 1926). British biochemist, Oxford, Sheffield.

Rathenau, Walther (1867–1922). German industrialist, philosopher, and politician.
Rehn, Eduard (1880–1972). Professor of Surgery, Freiburg University.
Renshaw, Alan. Technician in my team.
Ribbentrop, Joachim von (1893–1946).
Robbins, (Lord) Lionel Charles (b. 1898). British economist, Oxford, London.
Robb-Smith, Alistair. British pathologist, Oxford.
Robinson, Sir Robert (1886–1975). British chemist, Sydney, Liverpool, St Andrews, Manchester, Oxford.
Rona, Peter (1871–1945). Hungarian/German biochemist, Berlin.
Ross, Brian David (b. 1938). British medical scientist. Member of my team.
Ross, Sir Edward Denison (1871–1940). Oriental linguist, Professor of Persian, London.
Roughton, Geoffrey (1899–1972). British physiologist, Cambridge.
Rutherford, (Lord) Ernst (1871–1937). British physicist, Montreal, Manchester, Cambridge.

Salaman, Redcliffe Nathan (1874–1955). British biologist, London, Cambridge.
Sandford, Sir Folliott (Herbert) (b. 1906). Registrar, Oxford University.
Sanger, Frederick (b. 1918). British biochemist, Cambridge.
Sauerbruch, Ferdinand (1875–1951). German surgeon, Zurich, Munich, Berlin.
Sayer, Mr. British teacher at Technical College, Wembley.
Scheidemann, Philipp (1865–1939). German politician.
Schiller, Friedrich (1759–1805). German poet.
Schmidt, Friedrich, Nazi journalistic writer.
Schmitt, Francis Otto (b. 1903). US biologist, Dahlem, Massachusetts, Washington University (St Louis), Harvard, Massachusetts Institute of Technology.

Schönheimer, Rudolf (1898–1941). German/US biochemist, Freiburg, New York.
Shemin, David (b. 1911). US biochemist, New York, Chicago.
Siegfried II. Archbishop of Mainz from 1200–30.
Sies, Helmut (b. 1941). German biochemist, Munich, Düsseldorf.
Simpson, Esther. Secretary to Academic Assistance Council, (later The Society for the Protection of Science and Learning).
Sinclair, Sir Archibald (Viscount Thurso) (1890–1970). Leader of the Parliamentary Liberal Party, Secretary of State for Scotland, Secretary of State for Air.
Sinclair, Hugh MacDonald (b. 1910). British biochemist, Oxford.
Slotin, I. US biochemist, Chicago.
Smiley, Norman B. British industrialist, London.
Smith, R. A. US biochemist.
Soames, Mary. Daughter of Sir Winston Churchill.
Söling, Hans-Dieter (b. 1929). German biochemist, Freiburg, Göttingen.
Solzhenitsyn, Alexander Isayevitch (b. 1918). Russian novelist.
Speer, Albert (b. 1905). Nazi Minister of Armament.
Spemann, Hans (1869–1941). German zoologist, Freiburg.
Spicer, Arnold. Czech/British industrial chemist.
Srb, Adrian Morris (b. 1917). Stanford, Cornell.
Srere, Paul Arnold (b. 1925). US biochemist, Ann Arbor, Dallas, Student of Lipmann.
Staudinger, Hermann (1881–1965). German chemist, Zürich, Freiburg.
Staudinger, Magda. Wife of Hermann Staudinger.
Stephenson, Marjory (1885–1948). British biochemist, Cambridge.
Stern, Curt (b. 1902). German/US geneticist, Berlin, Berkeley, California.
Stern, Erich (b. 1900). German/US physician, Hildesheim, Essen, Decatur (Illinois). Friend from childhood days.
Stevens, Sir Roger (Bentham) (b. 1906). British diplomat. Vice-Chancellor of Leeds University.
Stickland, Leonard. British biochemist, Cambridge, Leeds.
Stocken, Lloyd (b. 1912). British biochemist, Oxford.
Stoecker, Adolf (1838–1909). German theologian and politician.
Stubbs, Marion (Spry) (b. 1942). British biochemist. Member of my team at Oxford.
Stumpf, Paul Karl (b. 1919). US biochemist, Columbia, Michigan, Berkeley, Davis (California).
Svenson-Piehl, Berta. Swedish artist.
Szent-Györgyi, Albert (b. 1893). Hungarian/US biochemist, Szeged, Budapest, Woods Hole.

PEOPLE IN THE BOOK

Telemann, Georg Philipp (1681–1767). German composer.
Thannhauser, Siegfried (1885–1962). German physician, Munich, Heidelberg, Freiburg, Boston.
Theorell, Hugo A. T. (b. 1903). Swedish biochemist, Stockholm.
Thompson, Sir Harold ('Tommy') (b. 1908). British physical chemist, Oxford.
Thomson, Joseph John (1856–1940). British physicist, Cambridge.
Thunberg, Torsten (1853–1952). Swedish physiologist, Lund.
Tidow, Georg (1900–78). German physician. My colleague at Altona.
Tisdale, Dr. Senior officer of the Rockefeller Foundation.
Traube, Isidor (1860–1943). German physical chemist, Berlin, Edinburgh.
Trautschold, Willhelm. German painter.
Trevelyan, Sir George (1876–1962). British historian, Cambridge.
Trowell, O. A. British cell biologist, Cambridge, M.R.C. Radiobiology Unit, Harwell.

Vaihinger, Hans (1852–1933). German philosopher, Halle.
Van Slyke, Donald Dexter (1883–1971) US biochemist, New York.
Vaughan, Dame Janet (Maria) (b. 1899). British medical scientist, London, Oxford. Principal, Somerville College.
Veale, Sir Douglas (1891–1973). Registrar, Oxford University.
Veech, Richard Lewis (Bud) (b. 1935). US biochemist, Boston, Harvard Medical School, Washington.
Vennesland, Birgit (b. 1913). Norwegian/US biochemist, Boston, Chicago, Berlin.

Wada, Mitsunori (b. 1896). Japanese biochemist, Tokyo, Obihiro, and Utsunomiya Universities.
Waksman, Selman Abraham (1888–1973). Russian/US microbiologist, Rutgers University (New Brunswick).
Wald, George (b. 1906). US biochemist, Harvard.
Walker, Ian. British biochemist, Birmingham, Oxford.
Warburg, Otto (1883–1970). German biochemist, Berlin.
Wayne, Sir Edward J. (b. 1902). British physician, Sheffield, Glasgow.
Wayne, Lady (Nancy). Wife of above.
Weaver, Warren (1894–1978). US mathematician, Senior Officer, Rockefeller Foundation.
Weber, Hans Hermann (1896–1974). German physiologist, Rostock, Münster, Königsberg, Tübingen, Heidelberg.
Weil-Malherbe, Hans (b. 1905). German-born US biochemist. My collaborator in Freiburg and Cambridge.
Weiss, Paul (b. 1898). Austrian/US biologist, New York.

Weizmann, Chaim (1874–1952). Russian-born biochemist and Zionist leader, Manchester, London, Rehovoth. First President of Israel.
Welt, Louis G. (1913–1974). US physician, Yale.
Werkman, Chester Hamlin (1893–1962). US microbiologist, Ames (Iowa).
Werth, Maria (see Maria Krebs).
Westenbrink, Professor (1901–1964). Dutch biochemist, Utrecht.
Wheare, Sir Kenneth (1907–1979). Professor of Government and Public Administration, Oxford.
Wheeler, Sir Mortimer (1890–1976). British archaeologist.
Whittam, Ronald (b. 1925). British physiologist, Sheffield, Cambridge, Oxford, Leicester.
Widdowson, Elsie May (b. 1906). British biochemist and nutritionist, London, Cambridge.
Wieland, Heinrich (1877–1957). German chemist, Freiburg, Munich.
Wieland, Theodor (b. 1913). German chemist, Heidelberg.
Wiggins, David (b. 1952). My technician at Oxford.
Wilkinson, Sir Denys (Haigh) (b. 1922). British nuclear physicist, Cambridge, Oxford, Brighton.
Williamson, Dermot (Derek) Hedley (b. 1929). British biochemist.
Willstätter, Richard (1872–1942). German chemist, Zürich, Berlin, Munich.
Wind, Franz. German medical scientist. Collaborator of Otto Warburg (1925/6).
Windaus, Adolf Otto Reinhold (1876–1959). German chemist, Göttingen.
Wittgenstein, Anneliese. German physician, Berlin.
Wöhler, Friedrich (1800–82). German chemist.
Wong, D. T. O. US biochemist.
Wood, Harland G. (b. 1907). US biochemist, Ames (Iowa), Minneapolis, Cleveland (Ohio).
Woods, Donald Devereux (1912–1964). British biochemist, Cambridge, London, Oxford.
Wright, Sir Norman (Charles) (1900–70). British nutritionist, New York, Washington, Ayr, Glasgow, Reading.

Yudkin, John (b. 1910). British nutritionist, Cambridge, London.

Zernike, Frits (1888–1966). Dutch physicist, Groningen.
Zöllner, Friedrich (Fritz) (b. 1901). German otolaryngologist, Freiburg.
Zuckmayer, Carl (1896–1977). German writer.

REFERENCES

Preface

(1) Shaw, G. B. S. *Sixteen self sketches.* Constable, London (1949).
(2) Goethe, W. von. *Bedeutung des Individuellen in vollst.* (letzter hand.) Vol. 59, p. 215. Cotta, Stuttgart and Tübingen (1942).
(3) Butler, Lord R. A. *The difficult art of autobiography.* Clarendon Press, Oxford (1968). (This is the Romanes Lecture of 1967 given in Oxford. It comments generally on many autobiographies.)
(4) Wethered, H. N. *The curious art of autobiography.* Christopher Johnson Publishers, London (1956).
(5) Burr, A. R. *The autobiography: a critical and comparative study.* Constable, London (1909).
(6) Matthews, W. *British autobiographies.* Berkeley, London (1955).
(7) Pascal, R. *Design and truth in autobiography.* Routledge and Kegan Paul, London (1960).
(8) Sloterdijk, P. *Literatur und Lebensfahrung, Autobiographien der zwanziger Jahre.* Carl Hanser Verlag, München (1978).

Chapter 1

(1) *Hildesheimer Hausinschriften und der figürliche Schmuck Hildesheimer Fachwerkhüser.* Helmke Verlag Werner Jäckh, Hildesheim. (Compiled by Hermann Schütte.)
(2) Del Monte, E. *Tour de monde.* Hachette, Paris (1889). Republished with German translation by Gebrüder Gerstenberg, Hildesheim, under the title *Une ville de temps jadis*, with introduction and translation by Walter Konrad (1974).
(3) Haffner, S. *Anmerkungen zu Hitler.* p. 197. Kindler Verlag, München (1978).
(4) Saunders, H. St. George *Royal Air Force 1939–1945.* Vol. III, p. 281. HMSO, London (1954).
(5) Quoted in Note (2).
(6) Plato. *Apology* 21, p. 82; and 29, p. 106. The Loeb Classical Library, Heinemann, London. G. P. Putnam's Sons, New York (1933).
(7) From a speech in the House of Commons on 22 July 1908, reported in *The Times* on the following day. (I owe this information to the Librarian of Balliol College, Mr Vincent Quinn.)

(8) *Iliad* xi, 514.
(9) Fisher, H. A. L. *A history of Europe*. Edward Arnold, London (1936).
(10) Gray (Viscount) Edward. *Twenty-five years 1892–1916*. Vol. 2, Chapter 20. Hodder and Stoughton, London (1925).

Chapter 2

(1) Masefield, J. *Speech at Sheffield University* (1946). Privately printed.
(2) Goethe, W. and Schiller, F. (1797). (See Xenien, Büchmann, G. *Geflügelte Worte*. 32nd edn p. 247. Haude & Spener, Berlin (1972).)
(3) Guttmann, W. and Meehan, P. *The great inflation*. Gordon and Cremosi, London (1976).
(4) Fergusson, A. *When money dies*. William Kimber, London (1975).
(5) Randall, A. Letter to *The Times*. 21 April (1972).
(6) Vaihinger, H. *Die Philosophie des Als Ob*. Ed. 7 and 8. Verlag Felix Meiner, Leipzig (1922).
(7) Friedrich, O. *Before the deluge: a portrait of Berlin in the 1920s*. Harper & Row, London (1974).
(8) Einstein, A. (See Büchmann, G., *Geflügelte Worte* 32nd edn p. 50. Haude & Spener, Berlin (1972).) (Old Testament, Vulgate translation, joining two passages.)

Chapter 3

(1) Warburg, O. *Schwermetalle als Wirkungsgruppen von Fermenten*, Verlag Dr Werner Saenger, Berlin (1946).
(2) Willstätter, R. *Aus meinem Leben*. Zweite Auflage, p. 200. Verlag Chemie, Weinheim (1958).
(3) Chargaff, E. Building the tower of babble. *Nature, Lond.* **248**, 776–9, 26 April (1974).
(4) Hopkins, F. G. *Skandinavisches Archive für Physiologie*. **19**, 33–59 (1926).
(5) Hopkins, F. G. Über die Notwendigkeit von Instituten für physiologische Chemie. *Münch. Med. Woch.* **73**, 1586–7 (1926).

Chapter 4

(1) Bartley, W., Kornberg, H. L., and Quayle, J. R. (eds) (1970). *Essays in cell metabolism* Wiley-Interscience, New York (1970).

REFERENCES

(2) Hopkins, F. G. Presidential Address to Royal Society, delivered on 30 November 1932 and published in *Nature, Lond.* 10 December 1932.
(3) Cohen, P. P. and Hayano, M. *J. Biol. Chem.* **166**, 239 (1946).

Chapter 5

(1) Warburg, O. *Biochem. Z.* **142**, 68 (1923); **152**, 51 (1924).
(2) Marshall, E. K. *J. Biol. Chem.* **14**, 283 (1913); **15**, 487, 495 (1913). (*see also* Collen, G. E. and van Slyke, D. D. *J. Biol. Chem.* **19**, 211 (1914).
(3) Ringer, S. *J. Physiol.* **4**, 29, 222 (1883); **7**, 291 (1886).
(4) Locke, F. S. *Zbl. Physiol.* **8**, 166 (1894); **14**, 670 (1900); **15**, 490 (1901).
(5) Tyrode, M. J. *Arch. int. Pharmacodyn.* **20**, 205 (1910).
(6) Kossel, A. and Dakin, H. D. *Z. physiol. Chem.* **41**, 321; **42**, 181 (1904).
(7) Wada, M. *Biochem. Z.* **224**, 420 (1930).
(8) Ackermann, D. *Biochem. Z.* **203**, 66 (1931).
(9) Meyerhof, O. *Die chemischen Vorgänge im Muskel*. Verlag von Julius Springer, Berlin (1930).
(10) Cori, C. F. and Cori, G. T. *J. Biol. Chem.* **81**, 389 (1929).
(11) Cohen, P. P. and Hayano, M. *J. Biol. Chem.* **166**, 239 (1946).
(12) London, E. S., Alexandry, A. K., and Nedswedski, S. W. *Hoppe-Seyler's Z. physiol. Chem.* **227**, 5–6 (1934).
(13) Trowell, A. A. *J. Physiol.* **100**, 432 (1942).
(14) Bach, S. J. and Williamson, S. *Nature, Lond.* **150**, 575 (1942).
(15) Bronk, J. R. and Fisher, R. B. *Biochem. J.* **64**, 118 (1956).
(16) Srb, A. M. and Horowitz, N. H. *J. Biol. Chem.* **154**, 129 (1944).

Chapter 6

(1) Löffler, W. Zur Kenntnis der Leberfunktion unter experimentell pathologischen Bedingungen. *Biochem Z.* **112**, 164–87 (1920).
(2) Schneeberger, G. *Nachlese zu Heidegger*. Buchdruckerei Ag, Suhr, Berne (1962).
(3) Einstein, A. *The world as I see it*. Watts and Co., London (1940).
(4) Londonderry, Marquess of, *Ourselves and Germany*. Robert Hale Ltd, London (1938).

Chapter 7

(1) Edson, N. L., *Biochem. J.* **29**, 2082 (1935); **29**, 2498 (1935); **30**, 1855 (1936); **30**, 1862 (1936); **30**, 2319 (1936).
(2) Booth, V. *Writing a scientific paper*. The Biochemical Society, London (1975).
(3) Bernal, J. D. *The social functions of science*. London (1939).
(4) Crowther, J. G. *The social relations of science*. London (1941, revised 1967).

Chapter 8

(1) Acland, Sir Richard. Letter to the *Listener* 1 February (1973).
(2) Braunstein, A. E. and Kritzmann, M. G. *Enzymologia* **2**, 129 (1937).

Chapter 9

More detailed information can be found in standard textbooks on biochemistry and in [276] where references to the authors quoted are given.

(1) Thunberg, T. *Skand. Acad. Physiol.* **24**, 23 (1910).
(2) Quastel, J. H. and Wolldridge, W. R. *Biochem. J.* **22**, 689 (1928).
(3) Stern, J. R., Shapiro, B., Stadtman, E. R., and Ochoa, S. *J. Biol. Chem.* **193**, 703 (1951).
(4) Stern, J. R., Ochoa, S., and Lynen, F. *J. Biol. Chem.* **198**, 713 (1952).

Chapter 10

(1) Mellanby, K. *Human guinea pigs*. The Merlin Press, London (1945 and 1973).

Chapter 11

(1) Wood, H. G. and Werkman, C. H. *Biochem. J.* **30**, 48–53, 618–23 (1936).
(2) Wood, H. G. and Werkman, C. H. *Biochem. J.* **32**, 1262–71 (1938).
(3) Wood, H. G. and Werkman, C. H. *Biochem. J.* **34**, 7–14, 129–38 (1940).
(4) Wood, H. G., Werkman, C. H., Hemingway, A., and Nier, A. O. *J. Biol. Chem.* **135**, 789–90 (1940).

REFERENCES

(5) Örström, Å., Örström, M., Krebs, H. A., and Eggleston, L. V., *Biochem. J.* **33**, 995–9 (1939).
(6) Evans, E. A. Jr., *Biochem. J.* **34**, 829–37 (1940).
(7) Ruben, S. and Kamen, M. D., *Proc. Nat. Acad. Sci.* **26**, 418–22 (1940).
(8) Krebs, H. A. and Eggleston L. V. *Biochem J.* **34**, 1385–1395 (1940).
(9) Evans, E. A. Jr. and Slotin, I., *J. Biol. Chem.* **136**, 301–2 (1940) and *J. Biol. Chem.* **141**, 439–50 (1941).
(10) Vennesland, B. and Hanke, M. E., *J. Bac.* **39**, 139–69 (1940).
(11) Gladstone, G. P., Fildes, P., and Richardson, G. M., *J. Exptl. Path.* **16**, 335–48 (1935).
(12) Solomon, A. K., Vennesland, B., Klemperer, F. W., Buchanan, J. M., and Hastings, A. B. *J. Biol. Chem.* **140**, 171–82 (1941).

Chapter 12

(1) Summarized in *Biol. Rev.* **26**, 87 (1951).
(2) Mitchell, P. *Les prix Nobel 1977*, Stockholm (1978).

Chapter 13

(1) Warburg, O. *Schwermetalle als Wirkungsgruppen von Fermenten.* Verlag Dr Werner Saenger, Berlin (1946).
(2) Robb-Smith, A. H. T. *Lancet* **II**, 559 (1942).

Chapter 14

(1) Bentley, R. *Nature, Lond.* **276**, 673 (1978).
(2) Hirschmann, H. In *Comprehensive Biochem.* **48**, 39–65 (1974).
(3) Hanson, K. R. *J. Am. Chem. Soc.* **88**, 2731–42 (1966).
(4) Hirschmann, H. and Hanson, K. R. *Tetrahedron* **33**, 891–7 (1977).
(5) Ogston, A. G. *Nature, Lond.* **276**, 676 (1978).
(6) Cornforth, J. W. *Les prix Nobel en 1975* p. 121, Nobel Foundation, Stockholm (1976).

Chapter 15

(1) Zuckerman, H. *Scientific élite.* The Free Press, London (1977).
(2) Delbrück, M. Quoted by C. F. von Weizäcker (personal communication).
(3) Weiss, P. *Science, N.Y.* **101**, 101 (1945).
(4) Goethe, W. von *Maximen und Reflexionen* (*No. 410 Aus den Heften zur Morphologie*). Deutscher Taschenbuch Verlag, München (1963).

(5) Medawar, P. R. *The art of the soluble.* Methuen, London (1967).

Chapter 16

(1) Veale, D. in *The Oxford region: a scientific and historical survey* (ed. A. F. Martin and R. W. Steel) p. 183 (1954).

Chapter 17

(1) Harrod, R. F. *The Prof: a personal memoir of Lord Cherwell,* p. 55. Macmillan, London (1959).
(2) *Report of the Commission of Enquiry* (chaired by Lord Franks). Oxford University Press (1966).
(3) Wheare, K. C. *Government by Committee.* Oxford University Press (1955).
(4) Cornford, F. *Microcosmographia academica.* Bowes and Bowes, Cambridge. (First published 1908 and often reprinted.)
(5) Smith, R. A. and Gunsalus, I. C. *J. Am. Chem. Soc.* **76,** 5002 (1954).
(6) Wong, D. T. O. and Ajl, S. J. *J. Am. Chem. Soc.* **78,** 3230 (1956).
(7) Quayle, J. R. Microbial growth on C_1 compounds. *Process Biochemistry,* February 1969.
(8) Quayle, J. R. Microbial assimilation of C_1 compounds. *Biochem. Soc. Trans.* **8,** 1-10 (13th Ciba Medal Lecture) 27 June 1979.
(9) Walgate, R. Single cell protein organism improved. *Nature, Lond.* **284,** 503 (1980).
(10) King, P. P. In *Speaking of science* Collected lectures from the Royal Institution. Taylor & Francis, London (1977).
(11) Burton, K. *Biochem. J.* **62,** 315 (1956).
(12) Benzinger, T. H., Kitzinger, C., Hems, R., and Burton, K. *Biochem. J.* **71,** 400 (1959).
(13) Ruderman, N. B., Houghton, C. R. S., and Hems, R. *Biochem. J.* **124,** 639 (1971).
(14) Berry, M. N. and Friend, D. S. *J. Cell Biol.* **43,** 506 (1969).

References to biographical details of many scientists associated with biochemistry are to be found in Joseph S. Fruton, *Selected bibliography of biographical data for the history of biochemistry since 1800.* American Philosophical Society Library, Library Publication No. 6, Philadelphia, 1974.

List of Publications

This list includes papers which do not carry my name as an author; they report work which I initiated, closely supervised, and usually wrote for publication.

I consider the more important papers to be 34, 38, 41, 49, 52, 57, 66, 67, 74, 80, 83, 90, 120, 166, 167, 169, 182, 205, 216, 225, 226, 262, 325.

Essays of general interest are the following: 104, 153, 190, 221, 227, 228a, 276, 280, 290, 291, 294, 312, 321, 326, 328, 331, 346, 348, 349.

[1] Die Färbung des Skelettmuskels mit Anilinfarbstoffen. *Arch. mikroskp. Anat. [Entw Mech.]* **97**, 557 (1923).
[2] Zur Goldsolreaktion im Liquor cerebrospinalis. *Klin. Wschr.* **4**, 1309 (1925).
[3] Die Flockung des kolloidalen Goldes durch Eiweisskörper. *Biochem. Z.* **159**, 311 (1925).
[4] Die Theorie der Kolloidreaktionen im Liquor cerebrospinalis. *Z. ImmunForsch exp. Ther.* **44**, 75 (1925).
[5] Zur Theorie der Weichbrodtschen Sublimatreaktion im Liquor-cerebrospinalis. *Dt. med. Wschr.* **51**, 1771 (1925).
[6] Studien zur Permeabilität der Meningen, I–IV Mitoilung (with A. Wittgenstein). *Z. ges exp. Med.* **49**, 553 (1926).
[7] (With A. Wittgenstein.) Untersuchungen über die Permeabilität der Meningen. *Dt. med. Wschr.* **52**, 1161 (1926).
[8] (With A. Wittgenstein.) Die Abwanderung intravenös eingeführter Substanzen aus dem Blutplasma, I und II Mitteilung. *Pflügers Arch. ges. Physiol.* **212**, 268 (1926).
[9] (With A. Wittgenstein.) Ueber die Abwanderung intravenös eingefuhrter Farbstoffe aus dem Blutplasma. *Klin. Wsch.* **5**, 320 (1926).
[10] (With P. Rona.) Physikalisch-chemische Untersuchungen über die Isohämagglutination *Biochem. Z.* **169**, 266 (1926).
[11] (With D. Nachmansohn.) Vitalfärbung und Adsorption. *Biochem. Z.* **186**, 478 (1927).
[12] Ueber die Rolle der Schwermetalle bei der Autoxydation von Zuckerlösungen. *Biochem. Z.* **180**, 377 (1926).
[13] (With F. Kubowitz.) Ueber den Stoffwechsel von Carcinomzellen in Carcinomserum und Normalserum. *Biochem. Z.* **189**, 194 (1927).
[14] Ueber den Stoffwechsel der Netzhaut. *Biochem. Z.* **189**, 57 (1927).

[15] (With O. Warburg.) Ueber locker gebundenes Kupfer und Eisen im Blutserum. *Biochem. Z.* **190**, 143 (1927).
[16] Ueber das Kupfer im menschlichen Blutserum. *Klin. Wschr.* **7**, 584 (1928).
[17] Ueber die Wirkung von Kohlenoxyd und Licht auf Häminkatalysen. *Biochem. Z.* **193**, 347 (1928).
[18] Ueber die Wirkung von Kohlenoxyd und Blausäure auf Hämatinkatalysen. *Biochem. Z.* **204**, 322 (1929).
[19] Ueber die Wirkung der Schwermetalle auf die Autoxydation der Alkalisulfide und des Schwefelwasserstoffs. *Biochem. Z.* **204**, 343 (1929).
[20] Stoffwechsel der Zellen und Gewebe. In *Methodik der wissenschaftlichen Biologie.* Vol. 2. p. 1049 (1929).
[21] Ueber Hemmung einer Hämatinkatalyse durch Schwefelwasserstoff. *Biochem. Z.* **209**, 32 (1929).
[22] (With J. F. Donegan.) Manometrische Messung der Peptidspaltung. *Biochem. Z.* **210**, 7 (1929).
[23] Manometrische Messung der fermentativen Eiweissspaltung. *Biochem. Z.* **220**, 283 (1930).
[24] Versuche über die proteolytische Wirkung des Papains. *Biochem. Z.* **220**, 289 (1930).
[25] Manometrische Messung des Kohlensäuregehaltes von Gasgemischen. *Biochem. Z.* **220**, 250 (1930).
[26] Ueber die proteolytische Wirkung von Papain und Kathepsin. *Naturwissenschaften* **19**, 133 (1931).
[27] Ueber Aktivierung proteolytischer Fermente. *Naturwissenschaften* **18**, 736 (1930).
[28] Ueber die Wirkung der Monojodessigsäure auf den Zellstoffwechsel. *Biochem. Z.* **234**, 278 (1931).
[29] (With H. Rosenhagen.) Ueber den Stoffwechsel des Plexus Chorioideus. *Z. ges. Neurol. Psychiat.* **134**, 643 (1931).
[30] Manometrische Messung der Eiweissspaltung. In *Abderhaldens Handbuch d. biologischen Arbeitsmethoden.* Section IV, Part 1, p. 871 (1931).
[31] Ueber die Proteolyse der Tumoren. *Biochem. Z.* **238**, 174 (1931).
[32] (With K. Henseleit.) Untersuchungen über die Harnstoffbildung im Tierkörper. *Klin. W.* **11**, 757 (1932).
[33] (With K. Henseleit.) Untersuchungen über die Harnstoffbildung im Tierkörper II. *Klin. Wschr.* **11**, 1137 (1932).
[34] (With K. Henseleit.) Untersuchungen über die Harnstoffbildung im Tierkörper. *Hoppe-Seyler's Z. physiol. Chem.* **210**, 33 (1932).

[35] (H. Manderscheid.) Ueber die Harnstoffbildung bei den Wirbeltieren. *Biochem. Z.* **263**, 245 (1933).
[36] Ueber den Stoffwechsel der Aminosäuren im Tierkörper. *Klin. Wsch.* **11**, 1744 (1932).
[37] Untersuchungen über den Stoffwechsel der Aminosäuren im Tierkörper, *Hoppe-Seyler's Z. physiol. Chem.* **217**, 191 (1933).
[38] Weitere Untersuchungen über den Abbau der Aminosäuren im Tierkörper. *Hoppe-Seylers Z. physiol. Chem.* **218**, 157 (1933).
[39] (H. Westerkamp.) Ueber Ketosäuren im Blutserum. *Biochem. Z.* **263**, 239 (1933).
[40] (P. Ostern.) Methode zur Bestimmung von Oxalessigsäure. *Hoppe-Seylers Z. physiol. Chem.* **218**, 160 (1933).
[41] (With Th. Benzinger.) Ueber die Harnsäuresynthese im Vogelorganismus. *Klin. Wsch.* **12**, 1206 (1933).
[42] (With H. Eitel and A. Loeser.) Die Wirkung der thyreotropen Substanz des Hypophysenvorderlappens auf die Schilddrüse *in vitro*. *Klin. Wschr.* **12**, 615 (1933).
[43] Grösse der Atmung und Gärung in lebenden Zellen, Oppenheimers *Handb. Biochm.* 2. Auflage, Ergänzungswerk, I, 863 (1933).
[44] Abbau der Aminosäuren, *Oppenheimers Handbuch der Biochemie* 2. Auflage, Ergänzungswerk, I, 939 (1933).
[45] Abbau der Fettsäuren. *Oppenheimers Handbuch der Biochemie* 2. Auflage, Ergänzungswerk, I, 936 (1933).
[46] Atmung und Gärung in lebenden Zellen. *Tabul. Biol.* **III**, 209 (1933).
[47] Urea formation in the animal body. *Ergebn. Enzymforsch.* **III**, 247 (1934).
[48] Deamination of Amino-Acids. *Biochem. J.* **29**, 1620 (1935).
[49] The synthesis of glutamine from glutamic acid and ammonia, and the enzymic hydrolysis of glutamine in animal tissues. *Biochem. J.* **29**, 1951 (1935).
[50] (With H. Weil-Malherbe.) The Conversion of proline into glutamic acid in kidney. *Biochem. J.* **29**, 2077 (1935).
[51] (With N. L. Edson.) Micro-determination of uric acid. *Biochem. J.* **30**, 732 (1936).
[52] (With N. L. Edson and A. Model.) The synthesis of uric acid in the avian organism: hypoxanthine as an intermediary metabolite. *Biochem. J.* **30**, 1380 (1936).
[53] Intermediate metabolism of carbohydrates. *Nature, Lond.* **138**, 288 (1936).

[54] (With W. A. Johnson.) Metabolism of ketonic acids in animal tissues. *Biochem. J.* **31**, 645 (1937).
[55] Dismutation of Pyruvic Acid in Gonococcus and Staphylococcus. *Biochemical Journal* **31**, 661, (1937).
[56] Metabolism of amino acids and related substances. *Ann. Rev. Biochem.* **5**, 247 (1936).
[57] (With W. A. Johnson.) The role of citric acid in intermediate metabolism in animal tissues, *Enzymologia* **4**, 148 (1937).
[58] The intermediate metabolism of carbohydrates. *Lancet* ii, 736 (1937).
[59] (With W. A. Johnson.) Acetopyruvic acid ($\alpha\gamma$ -diketovaleric acid) as an intermediate metabolite in animal tissues. *Biochem. J.* **31**, 772 (1937).
[60] The role of fumarate in the respiration of *Bacterium Coli Commune*. *Biochem. J.* **31**, 2095 (1937).
[61] Micro-determination of α-ketoglutaric acid. *Biochem. J.* **32**, 108 (1938).
[62] (With E. Salvin and W. A. Johnson.) The formation of citric and α-ketogluteric acids in the mammalian body. *Biochem. J.* **32**, 113 (1938).
[63] Metabolism of amino acids and proteins. *A. Rev. Biochem.* **7**, 189 (1938).
[64] (With L. V. Eggleston.) The effect of insulin on oxidations in isolated muscle tissue. *Biochem. J.* **32**, 913 (1938).
[65] (With Å. Orström.) Microdetermination of mypoxanthine and xanthine. *Biochem. J.* **33**, 984 (1939).
[66] (With Å. Orström and M. Orström.) The formation of hypoxanthine in pigeon liver. *Biochem. J.* **33**, 990 (1939).
[67] (With Å. Örström, M. Örström, and L. B. Eggleston.) The synthesis of glutamine in pigeon liver. *Biochem. J.* **33**, 995 (1939).
[68] (W. A. Johnson.) Aconitase. *Biochem. J.* **33**, 1046 (1939).
[69] (P. P. Cohen.) Microdetermination of glutamic acid. *Biochem. J.* **33**, 551 (1939).
[70] (P. P. Cohen.) Transamination in pigeon breast muscle. *Biochem. J.* **33**, 1478 (1939).
[71] The oxidation of d(+)proline by D-amino acid oxidase. *Enzymologia.* **7**, 53 (1939).
[72] (With L. V. Eggleston.) Bacterial urea formation (metabolism of *Corynebacterium ureafaciens*). *Enzymologia* **7**, 310 (1939).
[73] (With P. P. Cohen.) Metabolism of α-ketoglutaric acid in animal tissues. *Biochem. J.* **33**, 1895 (1939).
[74] (With L. V. Eggleston.) The oxidation of pyruvate in pigeon breast muscle. *Biochem. J.* **34**, 442 (1940).

[75] The citric acid cycle. *Biochem. J.* **34**, 460 (1940).
[76] The citric acid cycle and the Szent–Györgyi cycle in pigeon breast muscle. *Biochem. J.* **34**, 775 (1940).
[77] (E. A. Evans, Jr.) The metabolism of pyruvate in pigeon liver. *Biochem. J.* **34**, 829 (1940).
[78] (With D. H. Smyth and E. A. Evans, Jr.) Determination of fumarate and malate in animal tissues. *Biochem. J.* **34**, 1041 (1940).
[79] (With L. V. Eggleston, A. Kleinzeller and D. H. Smyth.) The fate of oxaloacetate in animal tissues. *Biochem. J.* **34**, 1234 (1940).
[80] (With L. V. Eggleston.) Biological synthesis of oxaloacetic acid from pyruvic acid and carbon dioxide. *Biochem. J.* **34**, 1383 (1940).
[81] (A. Kleinzeller.) The effect of electrolytes on the respiration of pigeon breast muscle. *Biochem. J.* **34**, 1241 (1940).
[82] (D. H. Smyth.) Vitamin B_1 and the synthesis of oxaloacetate by *Staphylococcus*. *Biochem J.* **34**, 1958 (1940).
[83] Carbon dioxide assimilation in heterotrophic organisms. *Nature, Lond.* **147**, 560 (1941).
[84] (A. Kleinzeller.) The formation of succinic acid in yeast. **35**, 495 (1941).
[85] (K. Mellanby.) Digestibility of national wheatmeal. *Lancet* 14 March p. 319, (1942).
[86] (With D. M. Stephenson.) The utilisation of carbon dioxide by heterotrophic bacteria and animal tissues. *Rep. Prog. Chem.* **38**, 303 (1942).
[88] Urea formation in mammalian liver. *Biochem. J.* **36**, 758 (1942).
[89] (With M. M. Hafiz and L. V. Eggleston.) Indole formation in *Bacterium coli commune*. *Biochem. J.* **36**, 306 (1942).
[90] The Intermediary Stages in the Biological Oxidation of Carbohydrate. *Advances in Enzymology* **3**, 191 (1943).
[91] Urea synthesis in mammalian liver. *Nature, Lond.* **151**, 23 (1943).
[92] Carbon dioxide assimilation in heterotrophic organisms. *A. Rev. Biochem.* **12**, 529 (1943).
[93] (With L. V. Eggleston.) The effect of citrate on the rotation of the molybdate complexes of malate, citramalate and isocitrate. *Biochem. J.* **37**, 334 (1943).
[94] (With K. Mellanby.) The effect of national wheatmeal on the absorption of calcium. *Biochem. J.* **37**, 466 (1943).
[95] (A. Kleinzeller.) Oxidation of acetic acid in animal tissue. *Biochem. J.* **37**, 674 (1943).

[96] (With L. V. Eggleston.) Metabolism of acetoacetic acid in animal tissues. *Nature, Lond.* **154**, 209 (1944).

[97] (With L. V. Eggleston.) Micro-determination of *iso* Citric and *cis*-Aconitic acids in biological material. *Biochem. J.* **38**, 426 (1944).

[98] (With the Vitamin A sub-Committee.) Vitamin A deficiency and the requirements of human adults. *Nature, Lond.* **156**, 11 (1945).

[99] (With L. V. Eggleston.) Metabolism of acetoacetate in animal tissues. 1. *Biochem. J.* **39**, 408 (1945).

[100] (With J. C. Speakman.) The solubility of sulphonamides in relation to hydrogen-ion concentration. *Br. Med. J.* **i**, 47 (1946).

[102] (J. Tosic.) Oxidations in Acetobacter. *Biochem J.* **40**, 209 (1946).

[103] (With L. V. Eggleston and R. Hems.) Urea synthesis in mammalian liver. *Nature, Lond.* **159**, 808 (1947).

[104] Cyclic processes in living matter. *Enzymologia* **12**, 88 (1947).

[105] (With W. O. Sykes and W. C. Bartley.) Acetylation and deacetylation of sulphonamide drugs in animal tissues. *Biochem. J.* **41**, 622 (1947).

[106] The D- and L-amino acid oxidases. *Biochem. Soc. Symp.* **1**, (1948).

[107] (With L. V. Eggleston.) Metabolism of acetoacetate in animal tissues. 2. *Biochem. J.* **42**, 294 (1948).

[108] Vitamin-C requirement of human adults (with vitamin-C sub-committee) *Lancet.* 5 June, p. 853 (1948); and Medical Research Council Special Report Series No. 280 (1953).

[109] (With L. V. Eggleston.) Observations on transiminations in liver homogenates, *Biochem. Biophys. Acts.* **2**, 319 (1948).

[110] Quantitative determination of glutamine and glutamic acid. *Biochem. J.* **43**, 51 (1948).

[111] (With L. V. Eggleston and R. Hems.) Synthesis of glutamic acid in tissues. *Biochem. J.* **43**, 406 (1948).

[112] Inhibition of carbonic anhydrase by sulphonamides. *Biochem. J.* **43**, 525 (1948).

[113] (With F. J. W. Roughton.) Carbonic anhydrase as a tool in studying the mechanism of reactions involving H_2CO_3, CO_2 or HCO_3^-. *Biochem. J.* **43**, 550 (1948).

[114] (J. R. Stern.) Carbon dioxide fixation in animal tissues. *Biochem. J.* **43**, 616 (1948).

[115] (With W. A. Johnson.) Cell metabolism. *Tabulae Biologicae* **XIX**, 100 (1948).

[116] (With L. V. Eggleston and R. Hems.) Distribution of glutamine and glutamic acid in animal tissues. *Biochem. J.* **44,** 159 (1949).
[117] (With J. B. Biale.) Oxidative processes in minced flower buds of the cauliflower, Brassica oleracea. *Amer. J. Bot.* **35,** 806 (1948).
[118] (With J. R. Stern, L. V. Eggleston, and R. Hems.) Accumulation of glutamic acid in isolated brain tissue. *Biochem. J.* **44,** 410 (1949).
[119] (Withe A. E. Bender.) The oxidation of various synthetic α-amino acids by mammalian D-amino acid oxidase of cobra venom and the L- and D-amino acid oxidases Neurosporn orassa. *Biochem. J.* **46,** 210 (1950).
[120] The tricarboxylic acid cycle. *Harvey Lect.* Ser. XLIV, p. 165 (1950).
[121] (With E. M. Hume.) Vitamin A requirements of human adults. *Medical Research Council Report.* No. 264 (1949).
[122] Chemical composition of blood plasma and serum. *Ann. Rev. Biochem.* **19,** 409 (1950).
[123] Manometric determination of L-aspartic acid and L-asparagine. *Biochem. J.* **47,** 605 (1950).
[124] Body size and tissue respiration. *Biochem. Biophys. Acta* **4,** 249 (1950).
[125] (L. E. Hokin.) The synthesis and secretion of amylase by pigeon pancreas *in vitro. Biochem. J.* **48,** 320 (1951).
[126] Improved manometric fiuid. *Biochem. J.* **48,** 240 (1951).
[127] (With L. V. Eggleston and C. Terner.) *In vitro* measurements of the turnover rates of potassium in brain and retina. *Biochem. J.* **48,** 530 (1951).
[128] The measurement of the turnover rate of steady state systems in living tissues. *Radioisotope techniques,* Vol. I, Medical and physiological applications (Proceedings of the Isotope Techniques Conference, Oxford, 1951) HMSO, London.
[129] The use of 'CO_2-Buffers' in manometric measurements of cell metabolism. *Biochem. J.* **48,** 349 (1951).
[130] Urea synthesis. *The enzymes.* Vol. 2, p. 866. Academic Press Inc., New York (1951).
[131] (L. E. Hokin.) Amino-acid requirements of amylase synthesis by pigeon-pancreas slices. *Biochem. J.* **50,** 216 (1951).
[132] (P. M. Nossal.) Estimation of L-malate and fumarate by malic decarboxylase of *Lactobacillus arabinosus. Biochem. J.* **50,** 349 (1951).
[133] Neuere Entwicklungen auf dem Gebiete der Patho-Physiologie der Niere. *Verhandlungen der Deutschen Gesell-*

schaft für innere Medizin. Kongress für innere Medizin **58**, 113 (1952).

[134] (With R. E. Davies.) Biochemical aspects of the transport of ions by nervous tissue. *Biochem. Soc. Symp.* **8**, 77 (1952).

[135] (With H. L. Kornberg.) Carbon dioxide exchanges between tissues and body fluids. *Brit. Med. Bull.* **8**, 206 (1952).

[136] (P. M. Nossal.) The effects of glucose and potassium on the metabolism of pyruvate in *Lactobacillus arabinosus*. *Biochem. J.* **5**, 591 (1952).

[137] (With S. Gurin and L. V. Eggleston.) The pathway of oxidation of acetate in baker's yeast. *Biochem. J.* **51**, 614 (1952).

[138] (L. V. Eggleston and R. Hems.) Separation of adenosine phosphates by paper chromatography and the equilibrium constant of the myokinase system. *Biochem. J.* **52**, 156 (1952).

[139] (With Olga Holzach.) The conversion of citrate into *cis*-aconitate and *iso*-citrate in the presence of aconitase. *Biochem. J.* **52**, 527 (1952).

[140] (With R. Hems.) Some reactions of adenosine and inosine phosphates in animal tissues. *Biochim. Biophya. Acts.* **12**, 172 (1953).

[141] Some aspects of energy transformations in living matter. *Br. Med. Bull.* **9**, 97 (1953).

[142] The equilibrium constants of the fumarase and aconitase systems. *Biochem. J.* **54**, 78 (1953).

[143] Equilibria in transamination systems. *Biochem. J.* **54**, 82 (1953).

[145] (With K. Burton.) The free-energy changes associated with the individual steps of the tricarboxylic acid cycle, glycolysis and alcohol fermentation and with the hydrolysis of the pyrophosphate groups of adenosine triphosphate. *Biochem. J.* **54**, 94 (1953).

[146] (With A. Ruffo, Monica Johnson, L. V. Eggleston, and R. Hems.) Oxidative phosphorylation. *Biochem. J.* **54**, 107 (1953).

[147] (With Monica Johnson, M. A. G. Kaye, and R. Hems.) Enzymic hydrolysis of adenosine phosphates by cobra venom. *Biochem. J.* **54**, 625 (1953).

[148] (R. Hems and W. Bartley.) Preparation of ^{32}P-labelled adenosine triphosphate. *Biochem. J.* **55**, 434 (1953).

[149] The Sheffield experiment on the vitamin C requirement of human adults. *Proc. Nutr. Soc.* **12**, 237 (1953).

[150] (With W. Bartley and R. E. Davies.) Active Transport in animal tissues and subcellular particles. *Proc. R. Soc.* **B142**, 187 (1954).

[151] Considerations concerning the pathways of synthesis in living matter: synthesis of glycogen from non-carbohydrate precursors. *Bull. Johns Hopkins Hosp.* **95**, 19 (1954).

[152] Some aspects of the metabolism of adenosine phosphates. *Bull. Johns Hopkins Hosp.* **95**, 34 (1954).

[153] Excursion into the borderland of biochemistry and philosophy. *Bull. Johns Hopkins Hosp.* **95**, 45 (1954).

[154] (R. Whittam and R. E. Davies.) Energy requirements for ion transport in steady-state system. *Nature, Lond.* **173**, 494 (1954).

[155] (L. V. Eggleston and D. H. Williamson.) The turnover rates of the phosphate groups of flavin-adenine dinucleotide and adenosine triphosphate during oxidative phosphorylation. *Biochem. J.* **56**, 250 (1954).

[156] (W. Bartley.) The formation of phosphopyruvate by washed suspensions of sheep kidney particles. *Biochem. J.* **56**, 387 (1954).

[157] Energy production in animal tissues and in micro-organisms. In *Cellular metabolism and infections* (ed. E. Racker). Academic Press Inc., New York (1954).

[158] The citric acid cycle. Nobel Lecture, Les Prix Nobel en 1953. Stockholm 1954.

[159] (With L. V. Eggleston and V. A. Knovett.) Arsenolysis and phosphorolysis of citrulline in mammalian liver. *Biochem. J.* **59**, 185 (1955).

[160] (With L. V. Eggleston.) Arsenolysis and phosphorolysis of citrulline. In *Biochemistry of nitrogen*, p. 496, Academia Scientiarum Fennica. Helsinki (1955).

[161] (With R. Hems.) Phosphate transfer reactions of adenosine and inosine nucleotides. *Biochem. J.* **61**, 435 (1955).

[162] Die Steuerung der Stoffwechselvorgänge. *Deut. Med. Woch.* **81**, 1 (1955).

[163] (K. F. Gey.) The concentration of glucose in rat tissues. *Biochem. J.* **64**, 145 (1956).

[164] The effects of extraneous agents on cell metabolism. *Ciba Found. Symp. on Ionizing Radiations and Cell Metabolism*, 1956, pp. 92–103.

[165] Biochemical concepts in medicine. *Oxf. Med. Sch. Gaz.* **9**, 1 (1957).

[166] (With H. L. Kornberg.) A survey of the energy transformations in living matter. *Ergebnisse der Physiologie biologischen Chemie und experimentellen Pharmakologie.* **49**, 212–98 (1957).

[167] (With H. L. Kornberg.) Synthesis of cell constituents from

[168] C$_2$-units by a modified tricarboxylic acid cycle. *Nature, Lond.* **179**, 988 (1957).
[168] (With R. Whittam and R. Hems.) Potassium uptake by *Alcaligenes faecalis*. *Biochem. J.* **66**, 53 (1957).
[169] Control of metabolic processes. *Endeavour.* **16**, 125 (1957).
[170] Inhibitors of energy-supplying reactions. In *Symp. Soc. Gen. Microbiol. No. VIII*, 104 (1958).
[171] (With P. K. Jensen and L. V. Eggleston.) Phosphorolysis of citrulline by mammalian liver: the effect of a bacterial activator. *Biochem. J.* **70**, 397 (1958).
[172] The regulation of metabolic processes. *Tex. rep. biol. med.* **17**, 16 (1959).
[173] Rate limiting factors in cell respiration. In *Ciba Found. Symp. on the Regulation of Cell Metabolism*, pp. 1–10. J. & A. Churchill Ltd, London (1959).
[174] (With L. V. Eggleston.) Permeability of *Escherichia coli* to ribose and ribose nucleotides. *Biochem. J.* **73**, 264 (1959).
[175] Biochemical aspects of ketosis. *Proc. r. Soc. Med.* **53**, 71 (1960).
[176] The cause of the specific dynamic action of foodstuffs. *Drug Research (Arzneim.-Forsch.)* **10**, 369 (1960).
[177] (With D. Bellamy.) The interconversion of glutamic and aspartic acid in respiring tissues. *Biochem. J.* **75**, 523 (1960).
[178] (With J. M. Lowenstein.) The tricarboxylic acid cycle. *Metabolic Pathways* (ed. D. M. Greenberg) Vol. I, pp. 129–203. Academic Press Inc., New York (1960).
[179] (With R. G. Kulka and L. V. Eggleston.) The reduction of acetoacetate to β-hydroxybutyrate in animal tissues. *Biochem. J.* **78**, 95 (1961).
[180] The biochemical lesion in ketosis. *Arch. Int. Med.* **107**, 51 (1961).
[181] (With L. V. Eggleston and A. d'Alessandro.) The effect of succinate and amytal on the reduction of acetoacetate in animal tissues. *Biochem. J.* **79**, 537 (1961).
[182] The physiological role of the ketone bodies. The Third Hopkins Memorial Lecture. *Biochem. J.* **80**, 225 (1961).
[183] (With J. R. Williamson.) Acetoacetate as fuel of respiration in the perfused rat heart. *Biochem. J.* **80**, 540 (1961).
[184] Biological reductions in complex systems. In Proceedings of the Robert A. Welch Foundation Conferences on Chemical Research **5**, 77 (1961).
[185] Control of cellular metabolism. In *The molecular control of cellular activity* (ed. J. M. Allen) pp. 279–96. McGraw-Hill (1962).

[186] (With R. Hems.) Further experiments on the potassium uptake by *Alcaligenes faecalis*. *Biochem. J.* **82**, 80 (1962).

[187] (With D. H. Williamson and J. Mellanby.) Enzymic determination of D(-)-β-hydroxybutyric acid and acetoacetate in Blood. *Biochem. J.* **82**, 90 (1962).

[188] (With J. Mellanby and D. H. Williamson.) The equilibrium constant of the β-hydroxybutyric-dehydrogenase system. *Biochem. J.* **82**, 96 (1962).

[189] (With L. V. Eggleston.) The effect of dinitrophenol and amytal on the reduction of acetoacetate in the presence of succinate. *Biochem. J.* **82**, 134 (1962).

[190] Enzyme activity and cellular structure. In *Horizons in biochemistry*, pp. 285–92. Academic Press Inc., New York, (1962).

[191] Control of metabolism. *Exptl. Eye Res.* **1**, 350 (1962).

[192] (With R. J. Haslam.) Substrate competition in the respiration of animal tissues. The metabolic interactions of pyruvate and α-oxogluterate in rat liver homogenates. *Biochem. J.* **86**, 432 (1963).

[193] (With D. A. H. Bennett, P. de Gasquet, T. Gascoyne, and T. Yoshida.) Renal gluconeogenesis. The effect of diet on the gluconeogenic capacity of rat kidney cortex slices. *Biochem. J.* **86**, 22 (1963).

[194] (With T. Yoshida.) Renal gluconeogenesis. 2. The gluconeogenic capacity of the kidney cortex of various species. *Biochem. J.* **89**, 398, (1963).

[195] (With T. Yoshida.) Renal gluconeogenesis. 3. Muscular exercise and gluconeogenesis. *Biochem. Z.* **338**, 241 (1963).

[196] (With R. Hems and T. Gascoyne.) Renal gluconeogenesis 4. Gluconeogenesis from substrate combinations. *Acta biol. med. germ.* **11**, 607 (1963).

[197] Concluding address at the international Symposium on Mécanismes de régulation des activités cellulaires chez les microorganismes. Colloques Internationaux du Centre National de la Recherche Scientifique, (1963).

[198] (With T. Yoshida.) Muscular exercise and gluconeogenesis. *Biochem. Z.* **338**, 241 (1963).

[199] Renal gluconeogenesis. In *Adv. Enzyme Regul.* **1**, 385 (1963).

[200] (With R. J. Haslam.) The metabolism of glutamate in homogenates and slices of brain cortex *Biochem. J.* **88**, 566 (1963).

[201] (R. Balazs and R. J. Haslam.) The role of aspartate aminotransferase in glutamate metabolism in brain. *Biochem. J.* **88**, 28P (1963).

[202] (With P. de Gasquet.) Inhibition of gluconeogenesis by α-oxo acids. *Biochem. J.* **90**, 149 (1964).
[203] (With E. A. Newsholme, R. Speake, T. Gascoyne, and P. Lund.) Some factors regulating the rate of gluconeogenesis in animal tissues. In *Adv. Enzyme Regul.* **2**, 71 (1964).
[204] The metabolic fate of amino acids. In: *Mammalian protein metabolism* (Ed. H. N. Munro and J. B. Allison) pp. 125–76. Academic Press, New York (1964).
[205] The Croonian lecture 1963. Gluconeogenesis. *Proc. R. Soc.* **B.159**, 545, (1964).
[206] (With C. Dierks and T. Gascoyne.) Carbohydrate synthesis from lactate in pigeon liver homogenates. *Biochem. J.* **93**, 112 (1964).
[207] (With W. Bartley, D. E. Griffiths, and L. A. Stocken.) Tissue preparations *in vitro*. In *Biochemisches Taschenbuch* (ed. H. M. Rauen) 2nd ed. Part II. Springer-Verlag, Berlin (1964).
[208] (With R. Hems.) Reduced nicotinamide-adenine dinucleotide as a rate-limiting factor in gluconeogenesis. *Biochem. J.* **93**, 623 (1964).
[209] (With R. N. Speake and R. Hems.) Acceleration of renal gluconeogenesis by ketone bodies and fatty acids. *Biochem. J.*, **94**, 712 (1965).
[210] Metabolic interrelations in animal tissues. *Proc. Robert A. Welch Foundation Conf.* **8**, 101 (1965).
[211] (With L. V. Eggleston.) The role of pyruvate kinase in the regulation of gluconeogenesis. *Biochem. J.* **94**, 3C (1965).
[212] (With M. Woodford.) Fructose 1,6-diphosphatase in striated muscle. *Biochem. J.* **94**, 436 (1965).
[213] (With P. Lund.) Formation of glucose from hexoses, pentoses, polyols and related substances in kidney cortex. *Biochem. J.* **98**, 210 (1966).
[214] (With W. Gevers.) The effects of adenine nucleotides on carbohydrate metabolism in pigeon-liver homogenates. *Biochem. J.* **98**, 720 (1966).
[215] (With R. Hems, M. J. Weidemann, and R. N. Speake.) The fate of isotopic carbon in kidney cortex synthesizing glucose from lactate. *Biochem. J.* **101**, 242 (1966).
[216] (With R. Hems, B. D. Ross, and M. N. Berry.) Gluconeogenesis in the perfused rat liver. *Biochem. J.* **101**, 284 (1966).
[217] (With B. M. Notton and R. Hems.) Gluconeogenesis in mouse-liver slices. *Biochem. J.* **101**, 607 (1966).
[218] The regulation of the release of ketone bodies by the liver. *In Adv. Enzyme Regul.* **4**, 339 (1966).

LIST OF PUBLICATIONS

[219] Bovine ketosis. *Vet. Rec.* **78**, 187 (1966).
[220] Die Ursachen der Ketonkörperanhäufung im tierischen Organismus. *Naturw. Rdsch. Braunschw.* **20**, 47 (1967).
[221] Theoretical concepts in biological sciences. In *Current aspects of biochemical energetics* p. 83. Academic Press Inc., New York (1966).
[222] (With T. Gascoyne and B. M. Notton.) Generation of extramitochondrial reducing power in gluconeogenesis. *Biochem. J.* **102**, 275 (1967).
[223] (With B. D. Ross and R. Hems.) The rate of gluconeogenesis from various precursors in the perfused rat liver. *Biochem. J.* **102**, 942 (1967).
[224] (W. Gevers.) The regulation of phosphoenolpyruvate synthesis in pigeon liver. *Biochem. J.* **103**, 141 (1967).
[225] (With D. H. Williamson and P. Lund.) The redox state of free nicotinamide-adenine dinucleotide in the cytoplasm and mitochondria of rat liver. *Biochem. J.* **103**, 514 (1967).
[226] (With J. M. Nishiitsutsuji-Uwo and B. D. Ross.) Metabolic activities of the isolated perfused rat kidney. *Biochem. J.* **103**, 852 (1967).
[227] The making of a scientist. *Nature, Lond.* **215**, 1244 (1967).
[228] Mitochondrial generation of reducing power. In *Biochemistry of mitochondria* p. 105. Academic Press and PWN, London and Warsaw (1967).
[228a] The biologist's and the chemist's approach to biochemical problems. In *Reflections on biological research* (ed. Giulio Gabbianu). Warren H. Green Inc., St. Louis, Missouri (1967).
[229] (With M. W. Bates and D. H. Williamson.) Evidence for the existence of an extramitochondrial pathway of acetoacetate synthesis in rat liver. *Biochem. J.* **104**, 59P (1967).
[230] (With B. D. Ross, R. Hems, and R. A. Freedland.) Carbohydrate metabolism of the perfused rat liver. *Biochem. J.* **105**, 869 (1967).
[240] (With J. M. Haslam.) The permeability of mitochondria to oxaloacetate and malate. *Biochem. J.* **107**, 659 (1968).
[241] (With G. D. Baird, K. G. Hibbitt, G. D. Hunter, P. Lund, and M. Stubbs.) Biochemical aspects of bovine ketosis. *Biochem. J.* **107**, 683 (1968).
[242] (With R. Hems and M. Stubbs.) Restricted permeability of rat liver for glutamate and succinate. *Biochem. J.* **107**, 807 (1968).
[243] (With D. H. Williamson and M. W. Bates.) Activity and intracellular distribution of enzymes of ketone-body metabolism in rat liver. *Biochem. J.* **108**, 353 (1968).

[244] (With T. Gascoyne.) The redox state of the nicotinamide-adenine dinucleotides in rat liver homogenates. *Biochem. J.* **108**, 513 (1968).

[245] The effects of ethanol on the metabolic activities of the liver. *Adv. Enzyme Regul.* **6**, 467 (1969).

[246] (With M. J. Weidemann.) Acceleration of gluconeogenesis from propionate by DL-carnitine in the rat kidney cortex. *Biochem. J.* **111**, 69 (1969).

[247] (With R. A. Freedland, R. Hems, and M. Stubbs.) Inhibition of hepatic gluconeogenesis by ethanol. *Biochem. J.* **112**, 117 (1969).

[248] (With M. J. Weidemann.) The fuel of respiration of rat kidney cortex. *Biochem. J.* **112**, 149 (1969).

[249] (With J. L. Corbett, R. H. Johnson, J. L. Walton, and D. H. Williamson.) The effect of exercise on blood ketone-body concentrations in athletes and untrained subjects. *J. Physiol.* **201**, 83P (1969).

[250] Renal carbohydrate and fatty acid metabolism. In *Renal transport and diuretics*, p. 1. Springer-Verlag, Berlin (1969).

[251] (With R. L. Veech.) Pyridine nucleotide interrelations. In *The energy level and metabolic control in mitochondria*, p. 329. Adriatica Editrice, Bari (1969).

[252] (With P. G. Wallace, R. Hems, and R. A. Freedland.) Rates of ketone-body formation in the perfused rat liver. *Biochem. J.* **112**, 595 (1969).

[253] (With M. J. Weidemann and D. A. Hems.) Effects of adenine nucleotides on renal metabolism. *Nephron* **6**, 282 (1969).

[254] (With D. H. Williamson, D. Veloso, and E. V. Ellington.) Changes in the concentrations of hepatic metabolites on administration of dihydroxyacetone or glycerol to starved rats and their relationship to the control of ketogenesis. *Biochem. J.* **114**, 575 (1969).

[255] (With L. V. Eggleston.) Strain differences in the activities of rat liver enzymes. *Biochem. J.* **114**, 877 (1969).

[256] (With M. J. Weidemann and D. A. Hems.) Effects of added adenine nucleotides on renal carbohydrate metabolism. *Biochem. J.* **115**, 1 (1969).

[257] (With D. J. C. Cunningham, M. Stubbs, and D. J. A. Jenkins.) Effect of ethanol on postexercise lactacidemia. *Israel J. Med. Sci.* **5**, 959 (1969).

[258] (With R. H. Johnson, J. L. Walton, and D. H. Williamson.) Metabolic fuels during and after severe exercise in athletes and non-athletes. *Lancet* ii, 452 (1969).

[259] (With R. L. Veech and L. V. Eggleston.) The redox state of

free nicotinamide-adenine dinucleotide phosphate in the cytoplasm of rat liver. *Biochem. J.* **115**, 609 (1969).

[260] (With R. L. Veech, L. Raijman, and K. Dalziel.) Disequilibrium in the triose phosphate isomerase system in rat liver. *Biochem. J.* **115**, 837 (1969).

[261] (With R. L. Veech.) Equilibrium relations between pyridine nucleotides and adenine nucleotides and their roles in the regulation of metabolic processes. *Adv. Enzyme Regul.* **7**, 397 (1969).

[262] (With R. L. Veech.) Interrelations between diphospho- and triphospho-pyridine nucleotides. In *Mitochondrial structure and function*, p. 101. Academic Press, London (1969).

[263] The role of equilibria in the regulation of metabolism. In *Current topics in cellular regulation*. Vol. 1, p. 45. Academic Press, New York (1969).

[264] (With R. H. Johnson, J. L. Walton, and D. H. Williamson.) Post-exercise ketosis. *Lancet* **ii**, 1383 (1969).

[265] (With J. T. Brosnan and D. H. Williamson.) Effects of ischaemia on metabolite concentrations in rat liver. *Biochem. J.* **117**, 91 (1970).

[266] (With M. J. Weidemann, R. Hems, D. L. Williams, and G. H. Spray.) Gluconeogenesis from propionate in kidney and liver of the vitamin B_{12}-deficient rat. *Biochem. J.* **117**, 177 (1970).

[267] (With R. L. Veech.) Regulation of the redox state of the pyridine nucleotides in rat liver. In *Pyridine nucleotide-dependent dehydrogenases*, p. 413. Springer-Verlag, Berlin (1970).

[268] (With R. L. Veech and L. Raijman.) Equilibrium relations between the cytoplasmic adenine nucleotide system and nicotinamide-adenine nucleotide system in rat liver. *Biochem. J.* **117**, 499 (1970).

[269] Rate control of the tricarboxylic acid cycle. *Adv. Enzyme Regul.* **8**, 335 (1970).

[270] (With J. R. Perkins.) The physiological role of liver alcohol dehydrogenase. *Biochem. J.* **118**, 635 (1970).

[271] (With D. H. Williamson, M. Stubbs, M. A. Page, H. P. Morris, and G. Weber.) Metabolism of renal tumors *in situ* and during ischemia. *Cancer Res.* **30**, 2049 (1970).

[272] (With R. Hems.) Fatty acid metabolism in the perfused rat liver. *Biochem. J.* **119**, 525 (1970).

[273] (With H. F. Woods and L. V. Eggleston.) The cause of hepatic accumulation of fructose-1-phosphate on fructose loading. *Biochem. J.* **119**, 501 (1970).

[274] Sir Archibald Garrod. *In Oxford Medicine*, p. 127. Sandford Publications, Oxford (1970).

[275] Intermediary metabolism of animal tissue between 1911 and 1969. In *British biochemistry past and present*, p. 123. Academic Press, New York (1970).

[276] The history of the tricarboxylic acid cycle. *Perspect. Biol. Med.* **14,** 154 (1970).

[277] (With D. H. Williamson, M. W. Bates, and M. A. Page.) Activities of enzymes involved in acetoacetate utilization in adult mammalian tissues. *Biochem. J.* **121,** 41 (1971).

[278] (With M. A. Page and D. H. Williamson.) Activities of enzymes of ketone-body utilization in brain and other tissues of suckling rats. *Biochem. J.* **121,** 49 (1971).

[279] Professor Otto Warburg. *Naturw. Rdsch.* **24,** 1 (1971).

[280] How the whole becomes more than the sum of the parts. *Perspect. Biol. Med.* **14,** 448 (1971).

[281] (With H. F. Woods.) Lactate production in the perfused rat liver. *Biochem. J.* **125,** 129 (1971).

[282] Reflections on the role of tryptophan derivatives on metabolic regulations. In *Metabolic effects of nicotinic acid and its derivatives*. p. 1115. Hans Huber Publishers, Bern (1971).

[283] Die Bedeutung der Grundlagenforschung für die Medizin. In *Biochemie und Klinik des Insulinmangels. 6. Symposion der Forschergruppe Diabetes 1970*, p. 1. Georg Thieme Verlag, Stuttgart (1971).

[284] (With D. H. Williamson, M. W. Bates, M. A. Page, and R. A. Hawkins.) The role of ketone bodies in caloric homeostasis. *Adv. Enzyme Regul.* **9,** 387 (1971).

[285] (With M. Stubbs and R. L. Veech.) Control of the redox state of the nicotinamide-adenine dinucleotide couple in rat liver cytoplasm. *Biochem. J.* **126,** 59 (1972).

[286] (With J. F. Biebuyck and P. Lund.) The protective effect of oleate on metabolic changes produced by halothane in rat liver. *Biochem. J.* **128,** 721 (1972).

[287] (With J. F. Biebuyck and P. Lund.) The effects of halothane (2-bromo-2-chloro-1, 1, 1-trifluoroethane) on glycolysis and biosynthetic processes of the isolated perfused rat liver. *Biochem. J.* **128,** 711 (1972).

[288] Otto Meyerhof's ancestry. In *Molecular Bioenergetics and Macromolecular Biochemistry*, p. 14. Springer-Verlag Berlin (1972).

[289] Der Pasteur-Effekt and und die Beziehungen zwischen Atmung und Gärung in lebenden Zellen. *Naturw. Rdsch.* **25,** 387 (1972).

[290] Some facts of life—biology and politics. *Perspect. Biol. Med.* **25**, 387 (1972).
[291] Wissenschaftliche Forschung in der heutigen Medizin. *Verh. dt. Ges. inn. Med.* **78**, 1 (1972).
[292] (With J. T. Hughes and D. Jerrome.) Ultrastructure of the avian retina. An anatomical study of the retina of the domestic pigeon (*Columba liva*) with particular reference to the distribution of mitochondria. *Exp. Eye Res.* **14**, 189 (1972).
[293] The Pasteur effect and the relations between respiration and fermentation. *Essays Biochem.* **8**, 1 (1972).
[294] Otto Heinrich Warburg. *Biogr. Mem. Fellows R. Soc.* **18**, 629 (1972).
[295] (With H. F. Woods.) The effect of glycerol and dihydroxyacetone on hepatic adenine nucleotides. *Biochem. J.* **132**, 55 (1973).
[296] The discovery of the ornithine cycle of urea synthesis. *Biochem. Educ.* **1**, 19 (1973).
[297] (With H. F. Woods.) Xylitol metabolism in the isolated perfused rat liver. *Biochem. J.* **134**, 437 (1973).
[298] (With N. Cornell, P. Lund, and R. Hems.) Acceleration of gluconeogenesis from lactate by lysine. *Biochem. J.* **134**, 671 (1973).
[299] (With R. Hems and P. Lund.) Accumulation of amino acids by the perfused rat liver in the presence of ethanol. *Biochem. J.* **134**, 697 (1973).
[300] (With R. Hems and P. Lund.) Some regulatory mechanisms in the synthesis of urea in the mammalian liver. *Adv. Enzyme Regul.* **11**, 361 (1973).
[301] Two letters by Wilhelm Conrad Röntgen. *Notes Rec. R. Soc. Lond.* **28**, 83 (1973).
[302] (With R. Hems and P. Lund.) Regulatory mechanisms in the synthesis of urea. In *Inborn errors of metabolism*, p. 201. Academic Press, London (1973).
[303] (With L. V. Eggleston.) Regulation of the pentose phosphate cycle. *Biochem. J.* **138**, 425 (1974).
[304] (With D. F. Wilson, M. Stubbs, R. L. Veech, and M. Erecinska.) Equilibrium relations between the oxidation-reduction reactions and the adenosine triphosphate synthesis in suspensions of isolated liver cells. *Biochem. J.* **140**, 57 (1974).
[305] (With L. V. Eggleston.) The regulation of the pentose phosphate cycle in rat liver. *Adv. Enzyme Regul.* **12**, 421 (1974).
[306] (With N. W. Cornell and P. Lund.) The effect of lysine on

gluconeogenesis from lactate in rat hepatocytes. *Biochem. J.* **142**, 327 (1974).

[307] Metabolic requirements of isolated organs. *Transplantn. Proc.* **6**, 237 (1974).

[308] (With N. W. Cornell, P. Lund, and R. Hems.) Some aspects of hepatic energy metabolism. In *Regulation of Hepatic Metabolism, Alfred Benzon Symposium VI*, p. 549. Munksgaard, Copenhagen (1974).

[309] (With N. W. Cornell, P. Lund, and R. Hems.) Isolated liver cells as experimental material. In *Regulation of Hepatic Metabolism, Alfred Benzon Symposium VI*, p. 726. Munksgaard, Copenhagen (1974).

[310] On the overuse and misuse of medication. *Executive Health XI*. (1974).

[311] The discovery of the carbon dioxide fixation in mammalian tissues. *Mol. Cell. Biochem.* **5**, 79 (1974).

[312] (With F. Lipmann.) Dahlem in the late nineteen twenties. In *Lipmann Symposium, Energy, Biosynthesis and Regulation in Molecular Biology*, p. 7. Walter de Gruyter, Berlin (1974).

[313] (With H. F. Woods and K. G. M. M. Alberti.) Hyperlactataemia and lactic acidosis. *Essays Med. Biochem.* **1**, 81 (1975).

[314] (With P. Vinay.) Regulation of renal ammonia production. *Med. Clins. N. Am.* **59**, 595 (1975).

[315] (With M. Stubbs.) The accumulation of aspartate in the presence of ethanol in rat liver. *Biochem. J.* **150**, 41 (1975).

[316] (With R. Hems, and P. Lund.) Rapid separation of isolated hepatocytes or similar tissue fragments for analysis of cell constituents. *Biochem. J.* **150**, 47 (1975).

[317] The role of chemical equilibria in organ function. *Adv. Enzyme Regul.* **13**, 449 (1975).

[318] (With M. Stubbs.) Factors controlling the rate of alcohol disposal by the liver. In *Alcohol intoxication and withdrawal*, p. 149, Plenum Publishing Corporation, New York (1975).

[319] (With J. H. Shelley (eds.).) The creative process in science and medicine. *International Congress Series*. No. 355. Excerpta Medica, Amsterdam (1975).

[320] (With P. Lund and N. W. Cornell.) Effect of adenosine on the ademine nucleotide content and metabolism of hepatocytes. *Biochem. J.* **152**, 593 (1975).

[321] The August Krogh Principle: 'For many problems there is an animal on which it can be most conveniently studied'. *J. Exp. Zoology*, **194**, 221 (1975).

[322] (With P. Lund and M. Stubbs.) Interrelations between glu-

coneogenesis and urea synthesis. In *Gluconeogenesis*, p. 269. Wiley, New York. (1976).

[323] (With R. Hems.) The regulation of the degradation of methionine and of the one-carbon units derived from histidine, serine and glycine. *Adv. Enzyme Regul.* **14,** 493.

[324] Concentration gradients between mitochondrial matrix and cytosol in the liver cell. In *Use of isolated liver cells and kidney tubules in metabolic studies*, p. 3. North-Holland Publishing Company, Amsterdam (1976).

[325] (With R. Hems and B. Tyler.) The regulation of folate and methionine metabolism. *Biochem. J.* **158,** 341 (1976).

[326] The discovery of the ornithine cycle. In *The urea cycle*, p. 1. Wiley, New York (1976).

[327] Soziologische Fragen der wissenschaftlichen Forschung. In *Vorlesungsreihe Schering*, p. 103. Pharma Forschung der Schering AG, Berlin (1976).

[328] Comments on the productivity of scientists. In *Reflections on biochemistry*, p. 415. Pergamon Press, Oxford (1976).

[329] *Festrede zur Feier des 750 jährigen Bestehens des Gymnasiums Andreanum Hildesheim, 29th November 1975*, p. 3. Gebrüder Gerstenberg, Hildesheim (1976).

[330] (With P. Lund.) Aspects of the regulation of the metabolism of branched-chain amino acids. *Adv. Enzyme Regul.* **15,** 375 (1977).

[331] Errors, false trails, and failure in research. In *Search and discovery—a tribute to Albert Szent-Györgyi*, p. 3. Academic Press, New York (1977).

[332] Getting to the root cause of delinquency. In *The Times.* 26 April, 1977.

[333] (With J. Viña and R. Hems.) Maintenance of glutathione content in isolated hepatocyates. *Biochem J.* **170,** 627 (1978).

[334] (With J. Viña and R. Hems.) Reaction of formiminoglutamate with liver glutamate dehydrogenase. *Biochem. J.* **170,** 711 (1978).

[335] (With P. Vinay and J. P. Mapes.) Fate of glutamine carbon in renal metabolism. *Am. J. Physiol.* **234,** F123 (1978).

[336] (With J. P. Mapes.) Rate-limiting factors in urate synthesis and gluconeogenesis in avian liver. *Biochem. J.* **172,** 193 (1978).

[337] (With M. Stubbs and P. V. Vignais.) Is the adenine nucleotide translocator rate-limiting for oxidative phosphorylation? *Biochem. J.* **172,** 333 (1978).

[338] Some general considerations concerning the use of carbohy-

drates in parenteral nutrition. In *Advances in parenteral nutrition*, p. 23. MTP Press (1978).

[339] Regulatory mechanisms in purine biosynthesis. *Adv. Enzyme Regul.* **16**, 409 (1978).

[340] (With D. Wiggins.) Phosphorylation of adenosine monophosphate in the mitochondrial matrix. *Biochem. J.* **174**, 297 (1978).

[341] Regulation of the concentration of low molecular constituents in compartments. In *Microenvironments and metabolic compartmentation*, p. 3. Academic Press, New York (1978).

[342] (With R. Hems, P. Lund, D. Halliday, and W. W. C. Read.) Sources of ammonia for mammalian urea synthesis. *Biochem. J.* **176**, 733 (1978).

[343] (With M. Watford and P. Lund.) Isolation and metabolic characteristics of rat and chicken enterocytes. *Biochem. J.* **178**, 589 (1979).

[344] (With G. M. Sainsbury, M. Stubbs, and R. Hems.) Loss of cell constituents from hepatocytes on centrifugation. *Biochem. J.* **180**, 685 (1979).

[345] Zur Biologie der Jugendkriminalität. *Wirtschaftspolitische Chronik* **1**, 27 (1979).

[346] On asking the right kind of question in biological research. In *Molecular mechanisms of biological recognition* p. 27 (ed. M. Balaban). Elsevier/North-Holland Biomedical Press, Amsterdam (1979).

[347] *Otto Warburg: Zellphysiologe, Biochemiker, Mediziner.* Wissenschaftliche Verlagsgesellschaft mbH, Stuttgart (1979).

[348] (With P. Lund and M. Edwards.) Criteria of metabolic competence of isolated hepatocytes. In *Cell populations: Methodological surveys* (B) *Biochemistry*: **9**, 1 (ed. E. Reid) Ellis Horwood Ltd. (1979).

[349] (With H. Taegtmeyer and R. Hems.) Utilization of energy-providing substrates in the isolated working rat heart. *Biochem. J.* **186**, 701 (1980).

[350] (With G. Baverel and P. Lund.) Effect of bicarbonate on glutamine metabolism. *Int. J. Biochem.* **12**, 60 (1980).

[351] Zur Biologie der Jugendkriminalität. *Sozialpädiatrie* **2**, 252 (1980).

[352] (With J. R. Krebs.) The 'August Krogh Principle'. *Comp. Biochem. Physiol.* **67B**, 379 (1980).

[353] Glutamine metabolism in the animal body. In *Glutamine: Metabolism, enzymology and regulation.* (ed. J. Mora and R. Palacios), 319. Academic Press (1980).

[354] Wie ich aus Deutschland vertrieben wurde. Dokumente

mit Kommentaren. *Medizinhistorisches Journal* **15**, 357. Gustav Fischer Verlag, Stuttgart, New York (1980).
[355] *Otto Warburg (1883–1970).* Oxford University Press (1981). (English translation of [347]).
[356] (With Jack E. Baldwin.) The evolution of metabolic cycles. *Nature, Lond.* **291**, 381 (1981).

INDEX

ATP 31, 47, 59
Abderhalden, E. 38
Abraham, E. P. Research Fund 240
Academic Assistance Council 89, 234–35
Acheson, R. M. 221
Ackermann, D. 57
Acland, Sir Richard 100
Adams, Sir Walter 234
Adrian, E. D. (Lord Adrian) 179
Agricultural Research Council 184, 242
Ajl, S. J. 218
Alder, K. 177
Altona, Municipal Hospital of 43–4, 46, 75
amino acids 44, 49, 54, 63, 86, 111, 114
Ammon, R. 25
Ampère, A. M. 153
Amsterdam 9, 23
Andreanum 9–11, 14
Anson, M. L. 38, 39
anti-semitism 5–6, 41, 61–4, 68–70, 74–8, 88, 154, 233
army service 14–15
Arthur Guinness, Son & Co. (Park Royal, Ltd) 204–5
Aschoff, E. 74
Aschoff, L. 20, 45, 73–4, 153
Asquith, H. H. 233
Attlee, C. R. 123
Aub, J. C. 38, 46, 87
Auburn (Auerbach), W. 44
Auerbach, W., *see* Auburn, W.
Austen, Canon A. 147–9
Austen, D. 147

Bach, S. J. 59, 92
Bacon, J. S. D. 135, 140, 187
Baeyer, A. von 176, 177
Baldwin, E. H. F. 86, 235
Balliol College, Oxford 84–6, 233
Barcroft, J. 64, 86, 179
Bartley, W. 135, 140, 145, 189, 219
Basle, University of 180
Beadle, G. W. 222
Beckman, A. O. 30
Beeson, P. 224, 225, 226
Bennett, G. M. 95
Benzinger, T. H. 45, 86, 219
Berlin 20, 22, 33, 42
Berlin-Charlottenburg, Technical University of 7
Berlin-Dahlem 25, 29–34, 39–40, 42, 48, 66, 80, 149, 168, 178, 195
Berlin, third medical clinic 22–4
Bernal, J. D. 90
Berry, M. N. 220
Berthollet, C. L. 153, 176
Beveridge, W. Lord 234–5
Bielschowsky, F. 45
Biochemical Journal v 135, 149
Biochimica et Biophysica Acta 149–50
Bismarck, O. von 81
Black Forest 18, 63
Blackett, P. M. S. [Lord] 166
Blaschko, H. H. 20, 31, 33, 83, 195
Bologna, University of 193
Booth, V. 88
Born, G. 243
Born, Max 243
Bosch works 157
Boston 38–9, 42, 66
Bowra, Sir Maurice 188
Bradley, S. 223
Bragg, Sir William (L) 208–9, 212

INDEX

brain drain 198
Braunstein, A. E. 103
Brighter Biochemistry 90
British Academy 240–2
British Biochemical Society v, 88, 91, 93, 149, 152, 231, 242, 245
Bronk, D. 39, 40, 59
Browning, R. 241
Bücher, T. 149, 150, 220
Buchner, E. 177
Buck, P. 181
Burk, D. 31, 151–2
Burton, K. 140, 142, 145, 189, 219
Butendandt, A. 152, 154, 177

Callow, A. B. 88
Cambridge, University of 20, 41, 59, 64–8, 79–80, 83–94, 96–7, 129, 152, 179, 187, 193–4, 195, 202, 206, 207–8, 231, 238, 244
cancer 23, 27, 52, 124
carbon dioxide fixation 103, 126–31
Carter, C. 84, 190
Cayman Islands (turtle farm) 98, 235–6
Cecil, R. 190, 221
cell respiration 28, 30, 36, 46–7, 52, 54
Chain, Sir Ernst 25, 92
Chamberlain, H. S. 6
Chamberlain, N. 100
Chance, B. 223
Chargaff, E. 34
Cherwell (1st Viscount), F. A. Lindemann 200
Cherreuil, M. E. 153
Chibnall, A. C. 237
Chicago, University of 129–30, 185, 222, 244
Chilver, G. A. F. 201
Christian, W. 30, 31
Churchill, Lady (Clementine Spencer) 168, 173
Churchill, Sir Winston 100, 168
citric acid cycle 50, 86, 97–8, 103, 105–18, 125, 138, 158–64, 215–18
Clark, G. A. 94, 95, 96
Coenzyme A 113–15, 216–18

Cohen, P. P. 49, 59, 103, 222
Cohen, R. H. C. 141, 144
Columbia University, New York 23, 223
Conant, J. B. 130
conscientious objectors 119–23, 133
copper 30, 37
Cori, C. F. 38, 58
Cori, G. T. 38, 58, 165
Cornford, F. 214–15, 240
Cornforth, J. W. 164
Correns, C. E. 31
Coward, Sir Noël 229
Cremer, W. 30
Crick, F. H. C. 208
Crowther, J. G. 90
Cuvier (Baron), L. C. F. D. (Georges) 153
cyclotrons 129–30
cytochrome oxidase 36

DNA 219
Dainton, Sir Frederick 223
Dakin, H. D. 55
Dalziel, K. 145, 146, 221
Daniel Sieff Institute 91
Davies, D. D. 223
Davies, R. E. 135, 138, 139, 140, 145, 167, 189
Davy, Sir Humphry 153
Delbrück, M. 177–8
Deuticke, H. J. 153
Deutsche Gesellschaft für Biologische Chemie 231, 246
dicarboxylic acids 86, 109–14, 118
Dickens, F. 190
Diels, O. P. H. 177
Dixon, G. 145
Dixon, K. C. 86
Dixon, M. 86
Donegan, J. F. 30
Donnan, F. G. 149
Donnan equilibria 23
Drury, Sir Alan 86
Dühring, C. E. 6

Ebert, F. 15
Edson, N. L. 86

INDEX

Eggleston, L. V.   97, 125, 139, 140, 189, 219, 226
Einstein, A.   24, 25, 74–5, 179
Eisenhower, D. D.   168
Eitel, H. (b. 1868)   62
Eitel, H. (b. 1902)   62
Ella Sachs Plotz Foundation, Boston   46
Elliott, K. A. C.   84
Elsevier company   150
Embden, G.   35
Embden–Meyerhof pathway   31
Emerson, R.   30, 38
England   37–8, 79, 83, 88, 154
enzymes   27, 28, 31, 36, 49, 59, 159–63, 212, 220
enzymologia   99
Epstein, F.   227
Estabrook, R. W.   231
Evans, E. A., Jr.   103, 117, 128–9, 159, 222
Exley, D.   189

Farrell, C. R.   140
fatty acids   19, 114, 220
Felix, K   153
Fieldhouse, M., *see* Krebs, M.
Fischer, E.   19
Fischer, E. H.   29, 175–7
Fisher, H. A. L.   13
Fisher, R. B.   60, 220
Florkin, M.   38
fluorides, exposure to   124
Foster, Sir Michael   179
Franks, (Baron) O. S.   202
'Franks Commission'   202–3, 214, 215, 239
Freiburg, University of   18–22, 35, 43–50, 62–73, 85, 168, 170, 244
Friedmann, E.   68, 92
Fujita, A.   30

Gaffron, H.   30
Gascoyne, A.   140
Gay-Lussac, J. L.   153, 176
Gebauer, J.   157
Geigy Company of Basle   225

Genevois, L.   31
Gerard, R.   31
Gibson, C. S.   234
gluconeogenesis   220
glutamine   63, 86, 103, 128, 138, 219
glyoxylate cycle   215–18, 240
Goldscheider, A.   22, 24
Goldschmidt, R.   31
Gollwitzer-Meier, K.   38, 42, 43, 152
Göttingen, University of   16, 18, 155–6, 236, 244
Government Food Policy Committee   120
Gray, C. T.   223
Green, D. E.   86, 88
Gregory, P.   140
Grey, Sir Edward   14
Guinness Fellowships   205; (*see also* Arthur Guinness Son & Co.)
Gunsalus, I. C.   218
Gurin, S.   223
Gustav Adolf, King of Sweden   173

Haas, E.   30, 31
Haber, F.   29, 195
Haddow, Sir Alexander   180
Hahn, O.   29
Haldane, J. B. S.   88
Hall, Sir Arthur   94
Hamburger, V.   31, 45
Hammarsten, E.   165
Hämmerling, J.   31
Handler, P.   222
Hanke, M.   130
Hanover   1, 14, 155, 244
Happold, F. C.   153
Harington, C.   93
Harris, L.   86
Harrison, D. C.   93
Harrod, Sir Roy   200
Hartmann, M.   31
Harvard University   38, 87, 129–30, 180–4, 203
Hastings, A. B.   129–30
Haworth, R. D.   185
Heidegger, M.   70–2

INDEX

Heidelberg, University of 32, 35, 48, 84
Hele, T. S. 96
Hems, R. 134, 139, 140, 189, 219, 226
Henseleit, K. 47, 82, 150
Herkel, W. 64
Herter lectures 185
Hess, B. 153
Hess, W. R. 102
Hevesy, G. de 45
Heyningen, W. E. van 185, 188, 190, 195
higher education, report on (Cmnd. 2154), *see* Robbins Committee
Hildesheim 1–9, 14, 80, 104, 147–9, 156–7, 194, 246
Hill, A. V. 102, 234
Hill, R. 86
Himsworth, Sir Harold 141–2, 187, 209, 211
histological staining 19
Hodgkin, D. M. C. 207–9
Holland 23, 99, 149
Holmes, B. 85, 86
Holmes, E. 86, 89
Home Office 124
honours 157, 171, 244–6; *see also* Nobel Prize
Hopkins, Sir Frederick Gowland
  Academic Assistance Council 234
  invitation to H. A. Krebs (1933) 65–6, 79, 84–5, 230
  Memorial Lecture (1960) v, 91–3
  offers of appointment to H. A. Krebs 87, 96
  personality 92, 235
  Presidential Address, International Physiological Congress (1932) 36
  Presidential Address, Royal Society (1932) 48, 64
Hoppe-Seyler, F. I. 36
Horowitz, N. H. 60
Huggins, C. B. 40
Hughes, D. E. 135, 138, 139, 140, 145, 189, 219
Hume, E. M. 121

hydrochloric acid production 139

Ilic, V. 140
Imperial Chemical Industries 219
inflation 20–1, 33
international physiological congresses 36, 38, 42, 66, 102, 152
iron porphyrins 28, 30, 37
isotopes, use of 116, 129–31, 159, 218
Israel 104; *see also* Palestine
Iveagh, (2nd Earl) 205
Iwasaki, K. 31

Jewish Refugees in Great Britain, Association of 240
Jews in Germany 5, *see also* anti-Semitism
Johns Hopkins University 36, 185
Johnson, W. A. 97–8
Jores, A. 44, 75–8
juvenile delinquency 227–8

Kaiser Wilhelm Gesellschaft 29, 35, 48
Kaiser Wilhelm Institute for Biology 25, 29–30, 40
Kamen, M. 129
Keilin, D. 28, 84–5, 86, 150–1
Kekulé, F. A. 176
Kempner, W. 30
Kendrew, Sir John 208
Kent, P. W. 220
Kenyon, Sir Frederic 234
ketone bodies 86, 138, 220
King, P. P. 219
Kistiakowsky, G. B. 130
Kitzinger, C. 219
Klingenberg, M. 220
Knoop, F. 19, 36, 38, 66–7, 110–11, 113
Kornberg, Sir Hans 135, 140, 142, 145, 189, 215, 223
Kossel, A. 35, 55
Kraepelin, E. 20
Krebs, A. 1, 6, 7, 18, 21
Krebs cycle, *see* citric acid cycle

INDEX

Krebs, E. 4, 9, 104
Krebs, Georg 1, 5, 6, 7, 8, 10, 12, 13, 22, 24, 80–2, 104
Krebs, Gisela 80, 81, 104, 147–8
Krebs, Helen 103, 171, 243
Krebs, J. 103, 171, 172, 243
Krebs, Margaret 89, 103, 166, 171, 184, 223, 231, 242–3
Krebs, Maria 80, 81, 104, 147–8, 156
Krebs, P. 103, 171, 243
Krebs, W. 4, 6, 7, 9, 13, 81, 104
Krehl, L. von 32
Kubowitz, F. 30, 31
Kuhn, R. 177

laboratory size 136–7
lactic acid production 28, 44, 52, 58, 107, 114
Lambert, R. A. 46, 67–8
Langley, J. N. 179
Laplace, Marquis de 153
Lascelles, J. 221
Laue, M. von 29
Lavoisier, A. L. 106, 176
Leaf, A. 227
League of Nations Technical Commission on Nutrition 121
Lehnartz, E. 153
Lemberg, R. 68, 89, 92
Lichtwitz, L. 43, 44, 75
Liebig, J. von 137, 176, 177
Lindemann, F. A., *see* Cherwell
Lindsay of Birker (1st Baron) 84
Lipmann, Freda 168
Lipmann, F. A. 25, 31, 33, 113, 167, 168, 177
Liverpool, University of 41, 129
Loeb, J. 39
Loeb, R. F. 39, 226
Löffler, W. 67, 79
Lohmann, K. 31
Lohmann reaction 31
London, E. S. 59
London, University of 180, 204, 205
London, University College 88, 93
Londonderry, (8th Marquess) 79

Lord Rank Research Centre 124
Lund, P. 134, 140, 219, 226, 227
Lunt, M. R. 221
Lwow, University of 48, 152
Lynen, F. 114, 177

McCance, R. A. 120
Mandelstam, J. 145
Mangold, O. 31
Manometry 28, 31, 38, 52, 87–8
Marshall, E. K. 52–3
Martius, C. 110–11, 113
Masefield, J. E. 16–17
Masson, Sir Irvine 132
'Matthew effect' 243
Max Planck Gesellschaft 29
Medawar, Sir Peter 177, 178
Medical Research Council 97, 120, 121, 132–4, 135, 139, 141, 143–4, 145, 181, 184, 186, 187, 189, 195, 208–9, 211, 212, 223, 226, 227, 237; *see also* MRC Unit
Medical Research Council Unit 132–46, 184, 186, 187–90, 195, 196, 199, 215, 218–20, 223
Meitner, L. 29
Mellanby, Sir Edward 132–3, 186, 237
Mellanby, K. 119, 122, 123
Mendel, B. 23, 24, 25, 29
Mendel, H. 25
Merton, R. K. 243
metabolic cycles (general) 47, 58
metabolic pathways 19, 31
Meyer, K. 25, 31
Meyer, O. 38
Meyerhof, O. 31–3, 35, 48, 58, 177
Michaelis, L. 35, 38, 39
Milligan, W. O. 222
Ministry of Food 124
Ministry of Health 119, 120, 124
Mitchell, P. 139
mitochondria 47, 108, 139
molecular biology (at Oxford) 207–12
Möllendorf, W. von 19, 45, 70
Moore, T. 86
Morton, R. A. 121

INDEX

Mosley, Sir Oswald 100
Müller, F. von 20
multi-professor departments 180, 203–7
Munich 9, 20
Munich Clinical School 20, 22

Nachmansohn, D. 23, 25, 31, 44, 91
National Socialism 5, 101; *see also* Nazism
Nature 98–9, 164, 212, 237
Nazism v, 1, 5, 7, 13, 47, 61–4, 68–79, 88, 92, 99, 102, 104, 147–8, 152, 153–7, 170, 240–1
Needham, D. M. M. 86
Needham, J. 86, 235
Negelein, E. 30, 31
Neuberg, C. 29, 48
New York 38, 76, 181
Newsholme, E. A. 140
nicotinamide 28
Niederdeutscher Beobachter 75–8
Nobel Foundation 154, 174
Nobel Prize (HAK) 41, 99, 165–179; *see also* Stockholm
Notgemeinschaft Deutscher Wissenschaft 80
Nottingham, University of 204, 223
Notton, B. M. 140
nutritional studies (wartime) 119–25

Oberstdorf 9
O'Brien, J. R. P. 84, 121, 195
Ochoa, S. 31, 33, 117, 177
Ogston, A. G. 117, 145, 146, 158–64, 190, 195, 220
Ord, M. 220
Ornithine cycle of urea synthesis 47–9, 51–60, 116, 118, 150, 230
Ostern, P. 48, 152
Osterwald, Dr 147
Öström, Å 103
Oxford 8, 84

Oxford, University of 12, 193–221, 239
 Annual Report 1962–3 207
 biochemistry at 41, 137, 138, 145, 195
 Commission of Inquiry (1966) 202–3
 Congregation 197, 200, 214
 Fellowships, *see* university salary scales
 Gazette 186, 199
 Hebdomadal Council 191, 198, 206, 213, 214, 215, 223–4, 242
 H. A. Krebs, appointment (1933) 67–8, 84–6
 H. A. Krebs, appointment (1954) 185–9
 H. A. Krebs, retirement 222–5
 Magazine 99, 201, 203, 239
 Royal Commissions on, 214
oxygen consumption 28, 52

Palestine 90–1
Paris, University of 23, 91
Parnas, J. 48, 152
Parsons, D. 220
Pasternak, C. 221
Pasteur effect 30
Pasteur, L. 164
Peacocke, A. R. 221
Pennsylvania, University of 145, 223
perfusion 220
Perutz, M. F. 207–8
Peter, A. 18
Peters, R. A. 67–8, 84–5, 121, 186, 188, 190, 195
Pflüger, E. 36
Phillips, Sir David 208–12
photosynthesis 27, 28
Pickering, Sir George 93, 195, 223–5, 237
Pirie, N. W. 86, 235
Pitts, R. F. 226
Planck, M. 48
Pohl, R. 18
Polanyi, M. 29
Pomerat, G. 181, 189

INDEX

Porter, R. R. 225
Prelog, V. 164
Pringle, J. W. S. 208–11
prochirality 236
promonta 76
Prussian Academy of Sciences 74–5

Quastel, J. H. 112
Quayle, J. R. 135, 140, 142, 145, 218–19

Radcliffe Infirmary, Oxford 135
Rathenau, W. 20, 21
Rehn, E. 50, 61, 170–1
Renshaw, A. 140, 190
retirement work 134, 222–9
'Robbins Committee' 201, 214
Robbins, Lord 201, 240
Robb-Smith, A. 153
Robinson, Sir Robert 188
Rockefeller Foundation v, 46, 66–8, 79, 83, 85, 87, 100–2, 143, 181, 184, 189, 196
Rockefeller Institute 36, 168
Rona, P. 24, 25
Ross, B. 234
Ross, Sir Edward 89, 235
Roughton, G. 86
Royal Cancer Hospital, London 180
Royal Institution 92, 208–9
Royal Society of London 23, 29, 64, 119, 142, 145–6, 149, 197
Rutherford, Lord 179, 234

Salaman, R. N. 90
Sandford, Sir Folliott 141, 224
Sanger, F. 208
Sauerbruch, F. 20
Sayer, Mr 37
scabies 119
Scheidemann, P. 15
Scheveningen 9
Schmidt, F. 76–8
Schmitt, F. O. 31
Schoenheimer, R. 45
school, *see* Andreanum

scurvy 122
Sheffield City Council 171
Sheffield, University of 16, 93–4, 95–104, 119, 132–5, 138, 139, 168, 181–5, 187–92, 194, 231, 244
Shermin, D. 163
Siegfried II, (Archbishop) 9
Sies, Helmut v, 231
Simpson, E. 234
Sinclair, Sir Archibald 100
Sinclair, H. McD. 186, 195, 237
Slotin, I. 129, 159
Smiley, N. B. 204–5
Smith, R. A. 218
Soames, M. 168
Söling, H.-D. 231
Solzhenitsyn, A. I. 99
Sorby Research Institute 119–23, 133
spectro photometry 28, 30
Speer, A. 4, 233
Spemann, H. 45
Spicer, A. 124
Srb, A. M. 60
Srere, P. A. 231
Staudinger, H. 45, 168, 170
Staudinger, M. 168
Stephenson, M. 86
Stern, C. 31
Stern, E. 16
Stevens, Sir Roger 172
Stickland, L. 86
Stocken, L. 190, 220
Stockholm, 165, 167, 168, 171–4
Stoecker, A. 6
Stubbs, M. 134, 140
Stumpf, P. K. 223
sulphonamides 204
Svenson-Piehl, B. 173
Szent-Györgyi, A. 38, 64–5, 108–10

Telemann, G. P. 10
Texas Christian University 222
Thannhauser, S. 43–7, 50, 66, 67, 150
Theorell, H. A. T. 167, 177

INDEX

Thomas Jefferson Research Center 228, 240
Thompson, Sir Harold 195
Thomson, J. J. 179
Thunberg, T. 112
Tidow, G. 44, 152
Times, The, of London 41, 235
Tisdale, Dr 101
tissue slice technique 28, 31, 46–7, 49, 52
Traube, I. 7
Traube's surface tension rule 7
Trautschold, W. 137
Trevelyan, Sir George 234
tricarboxylic acid cycle, *see* citric acid cycle
Trowell, O. A. 59
Tübingen, University of 35, 66

U-boat warfare 120
UNESCO 168
University Grants Committee 141–2, 196, 197, 198–200, 206, 211
university salary scales 142–5, 196–203, 211–12, 239
urea formation 47–9, 67, 86–7, 103; *see also* Ornithine cycle

Vaihinger, H. 21
Van Slyke, D. D. 38, 52
Vaughan, Dame Janet 190
Veale, Sir Douglas 187–8, 190
Veech, R. L. 226
Vennesland, B. 130
vitamin deficiency experiments 120–2, 195

Wada, M. 57
Waksman, S. A. 167
Wald, G. 180
Walker, I. 221
Warburg, O.
    appointment of H. A. Krebs 13, 25–6
    approach to Max Planck 48
    assessment of H. A. Krebs 40
    dismissal of H. A. Krebs 26, 40
    personality 27–9, 32–3, 150–2
    relationships with German universities 26, 35
    'scientific genealogy' 175–7
    under Nazis 66, 102, 149
    work of 27–8, 30, 36, 44, 53, 54
Wayne, Sir Edward 93–6, 100, 102
Wayne, Lady 94–5
Weaver, W. 101, 181
Weber, H. H. 25, 102, 152
Weil-Malherbe, H. 92
Weiss, P. 178
Weizmann, C. 90
Weizmann Institute of Science 91, 245
Wellcome Trust 196
Welt, L. G. 227
Werkman, C. H. 126–7, 129, 159
Westenbrink, Professor 150
Wheare, Sir Kenneth 213, 239
wheatmeal extraction 120
Wheeler, Sir Mortimer 240
Whittam, R. 135, 140, 145, 219
Widdowson, E. M. 120
Wieland, H. 150–1, 177
Wieland, T. 153
Wiggins, D. 140
Wilkinson, Sir Denys 201
Williamson, D. 134, 140, 190, 219, 220, 226, 227
Willstätter, R. 29, 150, 177
Wind, F. 30
Windaus, A. O. R. 18, 177
Wisconsin, University of 185, 222
Wittgenstein, A. 23, 24
Wöhler, F. 234
Wong, D. T. O. 218
Wood, H. G. 117, 126–7, 129–30, 159, 161
Woods, D. D. 86, 145, 190, 195, 204–5, 220
Woods Hole 38–9
World War, First 13–15, 16
World War, Second 95, 104, 119–25, 147, 231
Wright, Sir Norman 120